Wild Forest Lands

Wild Forest Lands

Finding History and Meaning in the Adirondacks

Philip G. Terrie

Syracuse University Press

For a listing of books published and distributed by Syracuse University Press,
visit https://press.syr.edu.

ISBN: 9780815611882 (paperback)
9780815657491 (e-book)

Library of Congress Cataloging in Publication Control Number: 2025014421

The authorized representative in the EU for product
safety and compliance is Mare Nostrum Group B.V.
Mauritskade 21D, 1091 GC Amsterdam, The Netherlands
gpsr@mare-nostrum.co.uk

Publication supported by a grant from

The Community Foundation *for* **Greater New Haven**

as part of the Urban Haven Project

For Greg Nobles,
Whose reliable and enduring friendship has meant a lot to me

Contents

Acknowledgments

Several people have read all or parts of earlier iterations of this book, and I am enormously grateful for their conscientious input. Their suggestions for correction, clarification, deletion, reorganization, and expansion have been invaluable. Peter Bauer, Dave Gibson, Pete Nelson, Claudia Braymer, Brad Edmondson, Phil Brown, Jackie Jablonski, and two reviewers for Syracuse University Press read with care. They bear no responsibility for any persisting errors. Special thanks are owed to Peter Bauer, whose energy and dedication were critical to Protect's success in the Article 14 litigation and who answered multiple emails and phone calls. Various librarians have steered me in the right direction and answered strange questions. For decades, Jerry Pepper at the Adirondack Museum fielded my phone calls, checked the shelves for me, and sent me photocopies and PDFs. After Jerry retired, Ivy Gocker and Jenny Ambrose did the same. The library at the Museum, now the Adirondack Experience, has for over sixty years been essential to my professional life. I am immeasurably grateful to Harold Hochschild for including a library in his original plans for the Museum and to all the directors and board members who have understood the value of this resource to the ever-growing community of researchers rummaging amid its treasures; I know many writers who share my appreciation. I was also helped at the Bowling Green State University Library, the Bancroft Library of the University of California, the Cornell University Law Library, the Tompkins County (New York) Library, the Ithaca College Library, and the New York State Archives, the latter with a Larry Hackman Research Residency Grant. A glance at my footnotes will show how important the *Adirondack Explorer* has been to my research. My life partner Jackie Jablonski tolerated stacks of books and papers on the dining room table and patiently endured it when my conversation relentlessly circled back to the wilderness and occasionally got lost in the blowdown. Emma took long walks with me while I pondered and sometimes solved organizational and other problems. Parts of this book were originally published, in different form, in the *Adirondack Explorer* and *Adirondack Life*. Finally, a tip of the cap to Syracuse University Press, with which I have had a long and successful relationship.

Wild Forest Lands

1

Topophilia

In 2017 the litigation known as *Protect the Adirondacks! Inc. v. New York State Department of Environmental Conservation and Adirondack Park Agency* began proceedings in the Supreme Court of Albany County. I served as an expert witness in this case, and this book builds on my involvement in what has turned out to be a monumentally important chapter in the history of the New York State Forest Preserve, most of which lies in twelve Adirondack counties. My role in the case derived from a career as a historian whose primary focus has been wilderness in the Adirondacks: why do we have it, what does it mean, how do we manage it, does it even exist? Preparing to testify, pondering the primary documents produced at the Forest Preserve's beginnings and throughout its subsequent evolution, and reading the mountains of paper piling up during the course of that litigation led me to rethink the entire subject, reassessing not only the origin story but also my own contributions to Adirondack historiography and the importance of wilderness to my personal, spiritual life.

In classical Greek, *topos* means place and *philia* derives from the root for love; "topophilia" means love of place. In 1974, geographer Yi-Fu Tuan published a book with this word as its title. Recommended to me in 1977 when I was starting the research for my dissertation on attitudes toward wilderness in the Adirondacks, it explores how love of place obtains in human lives.[1] What I am offering here is a book about my love for the Adirondacks, how it developed, what it means to me, how it has evolved over time, how it has shaped my professional and my inner life—how, with the exception of love for family and friends, it has meant more to me than any other feature of my adult life.

The fundamental feature of my love for the Adirondacks involves wilderness. Throughout my life and especially in the last few years, I have spent considerable energy and time contemplating how my understanding of wilderness, especially Adirondack wilderness, has morphed and grown. When we talk about wilderness in the Adirondacks, what do we mean? How do decades of scholarship about wilderness in American culture challenge facile assumptions, including some of my own initially unexamined understanding, about wilderness? This book asks: Is it reasonable to talk about wilderness in the Adirondacks? And if so, how do we do it?

The word "wilderness" (which I will use from here on without quotation marks, regardless of grammatical context) has come under intense scholarly scrutiny since I first hiked and camped in the Adirondacks. I will address this development in this book, but notwithstanding how it has often been deployed in a colonial, racially insensitive fashion or how its application to specific places is hotly contested and ineluctably ambiguous, I have decided to use it here. No other single word works.

Pondering wilderness in the Adirondacks inevitably leads to the New York Constitution, with its provision that the Forest Preserve must be "forever kept as wild forest lands" and that "the timber thereon shall not be sold, removed or destroyed." What, this book asks, does this mean?

Consideration of the constitution in turn leads us to the 2021 decision of the New York's highest court, the Court of Appeals, in *Protect the Adirondacks! Inc. v. New York State Department of Environmental Conservation and Adirondack Park Agency.*[2] This case (which will be identified throughout this book as *Protect*, in italics, to distinguish the litigation from the organization which initiated it, Protect the Adirondacks!, identified here by its widely recognized shorthand, Protect, typed without italics) began when the State proposed and then began building a network of road-sized snowmobile trails in the Forest Preserve. Protect believed and ultimately persuaded the New York Court of Appeals that these trails violated the state constitution. The litigation in *Protect*—the briefs filed and the decisions rendered—provides the backbone of this book. All the other chapters and excursions hither and yon are here because I'm trying to explain what led up to *Protect* and then to assess the implications. This book does not offer a comprehensive narrative of all the details of this case; rather it covers my own involvement and my understanding of its repercussions.[3]

It was while preparing for the first round of that trial, held at the Supreme Court in Albany,[4] and then again while trying to grasp the significance of the final decision handed down by the Court of Appeals, that I returned, some forty years or so after my first explorations, to the records of New York's constitutional conventions. I examined again the story of how Article 14 came to be and how its meaning has been such a contested point in Adirondack history. The fact that we have wilderness in the Adirondacks—whatever we may mean by invoking that ambiguous and often vaguely used word—is a function of the New York Constitution. So if we believe wilderness in the Adirondacks exists and is important, and if we defend it, we need to include the state constitution among our key documents. We must look carefully at the statutes that anticipated constitutional protection of the Forest Preserve; at the constitutional conventions of 1894, 1915, 1938, and 1967; and at two court decisions that preceded *Protect*, one in 1930 and one in 1991. All of these establish the context for the final decision in *Protect*. Pursuing this thread requires going through these documents with an attention to detail I have not previously deployed, looking especially at two key elements in the arguments of both sides in *Protect*: First, what has "timber" meant in all the relevant statutes and records? And second, what has "wild forest" meant in the same sources?

I am exploring a subject—wilderness in the Adirondacks—that has occupied me for over fifty years, running through many articles and four books. My scholarly record is largely a series of second thoughts and the need to refine what I have already written. At certain points in what follows, I refer to—while also trying to avoid repeating—what I have written elsewhere. My books have succeeded one another in the context of my evolving understanding of wilderness and its meanings; they thus constitute a parallel thread in my search for meaning in the

Adirondack wilderness, and I examine how they evolved, where they fell short, and how I set about amending the record.

Some people are fortunate enough to experience a deep love for the place where they were born and spend all their lives. My father was born in Charleston, West Virginia, in 1916, and, except for time at various schools and then in the Navy during World War II, he lived there his entire life. He practiced law in Charleston from the day he passed the West Virginia bar exam, as soon as he could manage after the end of the war, until, literally, two days before he died, in 1993. The culture, people, history, and all the other attributes of Charleston and West Virginia suited him just fine. He and my mother—who was born in Virginia, lived for a while in Eastern Kentucky, and moved to Charleston with her family as a child—liked to travel, especially once my sister and I were out of the house, but they never wanted to live anywhere else. Although they could have afforded one, they had no interest in a second or vacation home at the beach or on a mountain lake. Their ashes are buried at a hilltop cemetery with a grand view of Charleston and the Kanawha Valley.

Some of us, however fond we are of our birthplace and childhood roots—in my case also in Charleston—find our geographic lodestone somewhere else, while some never find it. Cultural critic and theorist bell hooks, whose good fortune it was to find, after many peripatetic years, that her home hearth was Kentucky, where she had begun, observes, "Many people feel no sense of place." This absence, consequently, leads to "a sense of crisis, of impending doom." She also notes, "When we love the earth, we are able to love ourselves." One of the recurring themes of her *Belonging* is her conviction that those blessed with a sense of place are those most likely to resist the relentless efforts of industrial capitalism to corrupt the places that provide meaning to those who love them.[5] One of the goals of this book is to acknowledge how lucky I feel for having found mine, relatively early in life.

What sort of book is this? It's a hybrid, mixing historical analysis, largely through close reading of primary and secondary documents, with personal experience. In both of these, the analysis and the memories, I am searching for meaning, and those two searches are intimately related. A recent example of this sort of combination is Ty Seidule's *Robert E. Lee and Me: A Southerner's Reckoning with the Myth of the Lost Cause.*[6] Like Seidule, I am trying to chart and understand my own journey through a subject of great importance to both my inner and my professional life.[7] Because the assessment of primary events is so intertwined with the story of the evolution of my thinking about wilderness, the chapters of this book are presented in a sequence reflecting that overlap.

Memory is notoriously unstable, particularly those memories we think we know best. As I understand it, when we remember an event from our past, we are recalling not the event itself but the last time we thought about it. So every time we recall what we think of as a well-established event, we're running it through yet another layer of the very act of recollection, and such filters have a cumulatively distorting effect. Those memories that we contemplate the most frequently are thus the ones most likely to be suspect. So when I reconstruct events that occurred decades ago, especially if certain details have become part of an often-repeated origin myth, it's possible or even likely I have it wrong.

Throughout this book, what I'm trying to figure out is *meaning.* What is the meaning—both rhetorical and spiritual—of Article 14, Section 1, of the New York Constitution? How does love of a place provide meaning for our lives? How, to be more specific, has the wilderness experience, rooted in my case in a specific place, the Adirondacks, provided meaning for me? How, moreover, did my obsession with Adirondack wilderness provide meaning to my life as a writer and scholar? How do all these intersect one another? What is the multiplier effect when they all come together?

What is my definition of wilderness? Is it a special place where body and soul are healed by the magical immanence of natural systems and the absence of obvious anthropogenic alterations? Is it a preserve set aside in the name of biodiversity? Is it a fictional construct, a projection of specific cultural and personal needs—often rooted in racially determined predispositions or the baggage of settler colonialism—onto physical reality? Is it a place where human action has disturbed the complex interconnections of non-human nature but where natural processes over time have largely erased the signs of that disturbance? In contemporary America, is it a place understood as such solely because of specific land-use management classifications, policies, and regulations, all of which depend on definitions of wilderness that are vague, imprecise, and subjective? Wilderness is all these things.

Often we understand what wilderness is only by defining it by the absence of an endless list of human intrusions, such as motorized vehicles, houses, factories, and shopping malls. In that case, at what point does wilderness become non-wilderness? If we hiked into the middle of the state-designated Five Ponds Wilderness Area in the northwest Adirondacks, for example, and found a bustling Burger King on the shore of the Oswegatchie River, I'm pretty sure we'd say that a place we thought was wilderness was no longer a wilderness. But what if we hiked in on a mostly well-marked trail, spent the night in a state-maintained lean-to, the shingles for which were flown in by a state-owned helicopter, and knew that we were camping in a state-designated "Wilderness Area"? Is it really a wilderness? Is the idea of a managed wilderness an oxymoron? All these questions raise important issues of degree. I hope the answers will come into focus in the following pages.

It may seem like a dodge, but defining wilderness, in any way other than as an administrative guide for what is and is not permitted, is ultimately a dead-end pursuit. Once we acknowledge that wilderness is a cultural construct, we accept that wilderness largely exists in the eye of the beholder, that it exists to the extent that it evokes powerful feelings of transcendence. It is a place where we *feel* and *do*, where the response is both transformative and performative.

It could be argued—and often has been—that there is no such thing as wilderness in the Adirondacks. Because so much of the Adirondacks was used and known by Indigenous cultures for centuries, if not millennia, before the arrival of Europeans with their obsessive attention to property titles and precisely marked boundaries, and because so much was logged, beginning in the nineteenth century and running right up to the present, some critics insist that the word simply has no reasonable application in the Adirondacks. Some say it has no application anywhere, that it's a projection of elite, western, and White cultural needs onto physical reality. Some scholars even treat the whole notion of wilderness and devotion to it with a barely disguised sneer, insisting that the "romantic idea of 'the wild' is born of human anxieties, particularly about urban and industrial life." A fixation with wilderness is an expression of "narcissism."[8] I understand this position, acknowledge its basis, and address it in this book. But, as we shall see, I believe the notion of wilderness has utility. I hope to explain why as we proceed.[9]

It's complicated—and endlessly fascinating. Writing twenty years before the critical constitutional provision of 1894, Verplanck Colvin, who ran the region's first extensive surveys, had this to say: "Few fully understand what the Adirondack wilderness really is. It is a mystery even to those who have crossed and recrossed it by boats along its avenues, the lakes; and on foot through its vast and silent recesses."[10] If it weren't mysterious, it would be a lot less interesting.

2

Contact

April 1966. I'm seventeen, at a boarding school in Virginia, a high-school senior, with a summer job lined up back in Charleston. A client of my father's, a businessman named Dempsey Jones, owns a warehouse from which he distributes various consumer goods. He's hired me to work there starting soon after graduation. Before I have a chance to find out what I'll be doing, Jones declares bankruptcy, and I no longer have a job waiting for me. For obvious and good reasons, my parents don't want me hanging around doing nothing all summer, and I'm instructed to score a job somewhere.

I happen to know that one of the teachers at Episcopal High School, where I was a student for grades ten through twelve, is on the staff at a camp somewhere in the Northeast, I'm not sure where. I was never in a class taught by Sandy Ainslie—or Mr. Ainslie as he was to me then—but for reasons I have forgotten I knew him well enough to ask if there might be a job for me at that camp. Two weeks after my initial inquiry, he summoned me to meet with him. I had been in and out of trouble that year, and my first thought was that I had been nailed for something and was getting kicked out with only a few weeks left before graduation. But, mercifully, Sandy was passing on the word that I could indeed work as a counselor at Adirondack Wilderness Camp (AWC), on Long Lake, in far upstate New York. My salary that first summer was $300.

My instructions were to get myself to Baltimore in mid-June. A chartered bus would drive me and about thirty or so other guys overnight to Long Lake. I don't recall even looking on a map to see where Long Lake was. I mentioned the name of the town to my father, and without giving it much thought he guessed maybe it was somewhere in the Finger Lakes. That seemed reasonable. Did neither of us know that the Finger Lakes, where I now live, and the Adirondacks are two entirely distinct—geologically, ecologically, historically, and culturally—parts of New York?

So I flew to Baltimore. The reason for assembling there was that the camp's director, Elliott Verner, taught at the Gilman School and had recruited most of his staff and campers from around Baltimore. The camp was staffed and attended largely by boys who attended or had recently graduated from private schools in Baltimore, with others from Wilmington, Delaware, where the camp's owner, Marianna Silliman, lived and had numerous contacts. It would be many years before I began to appreciate how much of my life, including my introduction to the Adirondacks, was a function of class and race privilege. I had just spent three years at all-White and all-boys Episcopal High School, an old and prestigious boarding school that prepared the sons of comfortably well-off, and in some cases fabulously wealthy, Southern families for college.[1] In September I would be a freshman at Princeton. I was an oblivious and lucky kid. At AWC I fit right in, culturally and socially, more than I knew or would have been self-aware enough to admit.

We arrived at the Long Lake town dock early on a June morning, having hit and killed a white-tailed deer on some twilit road shortly before sunrise. There we were met by Elliott, his brother Bill, and a few others. Bill gave us each a small bottle of Ole Time Woodsman's fly dope, and I thought, what the hell is this for? Then we were instructed to gather our duffels and climb into a fleet of small motorboats parked at the dock. Boats? No one had told me we were getting to camp by boat. I had been in a motorboat maybe twice in my life, and I've never been a strong swimmer.

That was my first trip from the town dock to the north end of Long Lake, a trip I have now made so many times I can visualize every foot of it—something I often do when lying in bed and waiting to fall asleep. I've navigated it in driving rain, in dense fog with a hand-held compass, frequently at night, and even a couple of times while it was snowing. I've walked, skied, and snowshoed on the lake when frozen and have driven rented cars on it (having opted not to tell the agent at the Albany airport where I was taking their car). On that early June morning, the lake was calm, a mirror, and the sun was shining. I was sleep deprived and anxiously wondering what in the world I had gotten myself into. Bug dope? Boats? It was about nine miles from the Long Lake town dock to AWC. Once we got to camp, I had my first encounter with blackflies, quickly figuring out what that bug dope was for. Mid-June: peak blackfly season.

But I settled in. It was a week before the campers would arrive, and we were put to work cleaning, painting, setting up the waterfront—all the routine chores needed to get an eighty-acre campsite with one of the best swimming beaches on the lake ready for nearly a hundred campers, also coming up by bus, mostly from Baltimore, with another faction from Wilmington, and from a handful of other locales around the Northeast. There were several other counselors from EHS. It was a good group. I liked the people I was working with, the work itself was not hard, and I was on a relatively remote shore of a spectacularly lovely fresh-water lake. I did not know it at the time, of course, but I was in the place, the north end of Long Lake, that would dominate my spiritual geography for the rest of my life.

I was also on the edge of the wilderness that would be the primary focus of an Adirondack obsession. From AWC, I could see the Sewards to the northeast, Santanonis to the east, and Kempshall to the south, all of which I would come to know well and in 1972 would be included in the state-designated High Peaks Wilderness Area. A few years later, I first encountered what Adirondack Murray had to say about precisely this view: "If you desire to see some of the best scenery imaginable, pass up the Raquette to Long Lake, and, when some two miles up the lake, turn your face toward the north, and you will behold what is worth the entire journey to see."[2] This is the view that greeted us every clear day at AWC.

3

Bill Verner

Toward the end of my first week at AWC, Bill Verner invited any interested counselors to hike with him to Round Pond for an overnight. AWC was built around one major premise: every boy should spend as much time as possible in the woods. The usual in-camp activities like riflery, baseball, and swimming were there to occupy the boys between trips, but the overarching aim—largely built around Bill's environmental, educational, and philosophical convictions—was getting them into the backcountry. Short trips for the younger kids, longer for the older. For every age group, the trips would get longer as the summer proceeded. Bill was in charge of planning the camping program for all age groups.

He already knew those counselors who had been at AWC the previous summer, its first. He thought we all needed a respite from the chores in camp, and I expect he wanted to pick up a sense of how comfortable the new guys like me were in the backcountry.

From 1959 through 1963 I had been a camper and then briefly an apprentice junior counselor at a camp on the Cowpasture River in Bath County, western Virginia. It was mostly a conventional camp: baseball, tennis, archery, target shooting with .22s, some riding, and wonderful hours-long games of capture the flag, played over an entire mountain and lasting until midnight. Unlike AWC, it seemed fixated on awards. There were winners and losers in every sport. And at the end of the summer, ribbons and trophies were distributed with abandon—but not to everyone, of course. At AWC, Elliott decreed that there would never be an award for anything. We worked, played, and camped in teams. Doing anything well was its own reward. Bill believed that navigating safely and amicably through the backcountry demanded teamwork and cooperation rather than individual accomplishment. Elliott especially wanted nothing to do with the medals and certificates with which the National Rifle Association had assiduously inserted its propaganda into the world of American camps.

At my Virginia camp, we always went on a few overnights each summer, camping in and around the George Washington National Forest. In Charleston, I had been a less-than-enthusiastic Boy Scout, not much interested in merit badges (never could learn the Morse Code), but more than willing to go out on camping trips. So before AWC, I had slept on the ground in a sleeping bag, could start a fire, could tie a few knots, and knew a white pine from a sugar maple. My father, who said he had spent far too many nights in muddy trenches on islands in the Pacific, had no interest in camping, and his family never hunted and seldom fished. My mother's father was a hunter with a well-locked and well-stocked gun closet and always a few bird dogs, but he died before I was old enough to be invited to spend any time in the woods with him.

Round Pond is a small lake, about a mile across and not especially round, in the northeastern corner of Hamilton County. From AWC we were ferried across Long Lake to Plumley Point, on the east shore, formerly the site of a small, family-run hotel that had closed only a few years before.[1] The few buildings that had constituted the hotel compound were gone, replaced by a state lean-to (there are now two). From Plumley's, we hiked east on the Northville-Lake Placid Trail for about two miles before veering off to the south on an unmarked, seldom-used, and hard-to-follow hunter's track to Round Pond, about a mile off the NPT.[2] It was an interesting night: the mosquitoes were ferocious. But I also had my first opportunity to get acquainted with Bill Verner.

I have had a first-class education. From first grade through graduate school, I have had superb, interesting, dedicated teachers—knowledgeable and enthusiastic. But not one of all the men and women who taught me and graded my papers had the impact on me that Bill Verner did. Sitting around a fire that night at Round Pond, I heard for the first time someone talk with passion and intelligence about wilderness, a subject I had previously thought about not at all. I was a few weeks short of my eighteenth birthday and impressionable. He was articulate, well read, and funny, never taking himself too seriously. He was a paradoxical combination of sophisticated tastes in classical music and wine with indifference to appearances (he nearly always wore ragged sweaters and cheap pants from a discount department store) and all popular culture except film. Most important, he was driven by a passionate dedication to wilderness and its history and to protecting what remained of it in the Adirondacks. He wore his Ivy League education and intellect lightly, but he defended wilderness with deeply felt intensity. The whole package was galvanizing.

Bill was a curator at the Adirondack Museum, where he had worked out an agreement with the director, Jerry Swinney, to take off two months in the summer to give him time at AWC. Mixed in with his passion for wilderness and the spiritual uplift it offered was a well-informed grasp of how all this fit into Adirondack history; he was one of a handful of people at that point who were experts.[3]

The events of the summer of 1966, when I began my discovery of the Adirondacks, played out in the spell that Bill quite unintentionally cast over me. I heard a lot of terms that meant little to me at first but would eventually form the foundation of my professional and spiritual life. I heard about the New York Constitution and its provision protecting the Forest Preserve as "wild forest lands." I listened to and began to participate in discussions of leave-no-trace and wilderness ethics. I heard about the Forty-Sixers. After almost six decades I have no idea what had stuck by the time August and the end of the camp season rolled around. But over the summer and over the years, as Bill and I became friends and as we discussed all these and so much else late into so many nights, it gradually came into focus.

From the late 1960s and through the establishment and early years of the Adirondack Park Agency, Bill was actively involved in all Adirondack environmental matters. Among many other expressions of his environmental energy was his tireless lobbying when objections to the Park Agency were coalescing in the legislature in 1971.[4] The peak of his activism came in 1978, when he was appointed by the Agency to be chair of its Citizens' Advisory Task Force on Open Space, which submitted recommendations to the Agency in April of 1980. These offered a comprehensive list of goals, all designed to examine the importance of protecting open space in the Adirondacks and suggest management guidelines. Historian Barbara McMartin interviewed George Nagel, who was on the Agency staff at the time and worked closely with Bill, and wrote that Nagel characterized Bill thus: "Urbane, generous, and gracious as a task force leader and the kind of farsighted planner who was capable of reflecting on such parkwide issues."[5]

Early in my developing relationship with Bill, when it was still in the mentor-student phase, he recommended Roderick Nash's *Wilderness and the American Mind*, and I bought a copy of the first edition—the book had been in print for only a year—at the Adirondack Museum bookstore in 1968.[6] This book became my bible. It reinforced Bill's powerful combination of wilderness enthusiasm with intellectual history. Nash laid out a marvelous and persuasive narrative of how American attitudes toward wilderness had evolved from fear and hostility to appreciation and a wish to protect it. I have much more to say on Nash and his understanding of how Americans have dealt with what they perceived to be the vast wilderness of their country. It would be several decades before I began to understand its shortcomings. Notwithstanding these, my copy of the first edition remains a treasured item on my Adirondack bookshelf. A valuable reference, it always reminds me of Bill, who died at the age of fifty-three in 1989.

4

Couchsachraga

Because Sandy Ainslie had been the key to my being hired and because he supervised the youngest kids, known as the Juniors, I began my first summer at AWC working in his unit. That was not a good match. It takes maturity to work successfully with elementary-school kids. I was impatient with them in camp and not a successful trip leader in the woods. I began lobbying to join the older kids and get into the woods on some serious backcountry trips. It was yet another example of my incredibly good fortune that summer that Elliott, Bill, and Sandy were all willing to let me migrate. Toward the middle of the summer, I joined a trip of four or five nights into isolated country on the Cold River. That trip group already had two experienced senior counselors, Don McMullen and Nelson Bolton. My luck continued; neither of them objected.

This was the mid 1960s, just before the boom in high-tech camping gear. We were using what now seems to have been primitive equipment: ancient canvas packs and bulky sleeping bags, boots from an Army-Navy store. We cooked our meals over wood fires. For shelter, we had waterproof (more or less) squares of canvas that we'd string between trees. No gas cooking stoves, no mosquito netting. Every year, we'd up the level of our gear. By the time the camp folded in 1971, I was wearing better boots, carrying a Kelty pack and a Holubar sleeping bag, and using a gas cookstove. A year after that I bought the first of several lightweight tents, with mosquito netting. To this day, the Adirondack bug that annoys me the most is the mosquito: black-flies have the good sense to go to sleep at night, and they don't make any noise.

One of the ironies of the wilderness-camping movement that began for me at Round Pond in 1966 and for many other Americans in that decade was the ever-increasing prominence of high-tech gear. Most of us were telling ourselves that we were following in the footsteps of Natty Bumppo, exploring the wilderness, guided by self-reliance and a dismissal of the seductions of modernity.[1] But all this was made more appealing and more comfortable by backpacks that got lighter, more capacious, and less uncomfortable, by sleeping bags that got warmer and stuffed into smaller sacks, and by portable stoves that got more efficient. By the early '70s, we were sleeping on miracles of foam pads and carrying tents that weighed only a few pounds, were genuinely mosquito proof, and collapsed to something not much bigger than a shoebox. The trek into the wilderness was an explicit statement of anti-modernism, but behind it lurked, for most of us, an enthusiastic embrace of the rapidly expanding consumer culture.[2] It was one of several ways, I eventually realized, that our wilderness experience was to a large extent performative.

I had no idea where Cold River was and probably didn't even look at a map until we were sitting around a campfire the first night. We had followed old tote roads leading off the Northville-Lake Placid Trail near Shattuck's Clearing and found ourselves camping at Cold River Flow, a mile or so from the site of Noah John Rondeau's hermitage. After dinner, Nelson asked who wanted to head off the next morning on a bushwhack up Couchsachraga and then over to Panther, two peaks in the Santanoni Range. This was 1966, before there were major herd paths in the Santanonis; there were certainly none on the southwest slopes. At that time the recommended route up Cooch (our shorthand for Couchsachraga) was from an abandoned logging road and then more or less straight up to the peak, the lowest of the Forty-Six at 3,820 feet. Virtually no one traversed the ridge from Cooch to Panther. At some point, before we started, Nelson, who had been around in the Adirondacks for some time, mentioned the word "blow-down." Like blackflies and Forty-Sixers, this was new to me. "The traverse of the ridge between Couchsachraga and Panther Peak," opined the then-current guidebook, "is a difficult trip, being complicated by heavy blowdown, and is not recommended."[3]

In November of 1950, a fierce windstorm swept across the northeastern Adirondacks. It knocked over the trees on hundreds of thousands of acres and was especially destructive in the Seward and Santanoni ranges. There, above around 2,500–3,000 feet, almost every standing tree had been toppled or stripped of its branches. Sixteen years later the upper slopes of Cooch and Panther and the ridge between them presented a thick matrix of fallen tree trunks, through which grew a tangled, dense mass of young spruces and balsam firs. With plenty of sunlight to promote rapid growth, the spruces and firs had by then been colonizing these slopes for about fifteen years. A striking feature of a blowdown area was the naked trunks of white birches, stripped by the 1950 storm and poking up starkly through the spruce. To make your way, you would be walking on fallen trees sometimes two or three feet off the ground, able to see only a few yards as you worked your way through the new growth, which was just high enough that it was often impossible to see any peak around you. Map and compass work were essential. After a season or two of bushwhacking in blowdown, I also found a pair of work gloves to be useful.

That Cooch was my first peak seems providential. The name "Couxaxrage," where the "x" is used to represent the sound of the Greek chi, first appears on a 1755 "Map of the Middle British Colonies in North America" compiled by cartographer Lewis Evans; it is written precisely over that part of northern New York we now identify as the Adirondack Park. A narrative written to accompany this map by colonial official Thomas Pownall and published in 1776 explains that the word was of Indian origin and "signifies the Dismal Wilderness or Habitation of Winter." Pownall further writes, "It is said to be a broken, unpracticable Tract."[4] Wilderness!

This day it was Nelson, two or three boys, and me. We left camp at daybreak and backtracked on the abandoned logging road to a deteriorating bridge built for trucks a few decades earlier. The guidebook designated this bridge as the "key" to climbing Cooch. We headed up the creek that passed under the bridge and started climbing. It was a classic Adirondack bushwhack, with seldom a trace, so far as I recall, of evidence of previous hikers, though we knew that at least some early Forty-Sixers had preceded us. Anyone who bagged peaks in the '60s will know what it was like: slow, hot, and tiring. But we were well into July, and the humidity was low that day, so the blackflies were minimal. I have no recollection of how long it took to reach the canister on Cooch. I'd guess two or three hours. The long trek over the ridge to Panther, where the blowdown was even denser, is another vague recollection. What I do recall, with cinematic clarity, is the view from Panther.

To say that something changed your life seems a cliché, but this was one of those days for me. It was a time of year when many sunny days in the Adirondacks are hot, humid, and hazy, with less-than-ideal views. But this was a perfect, cloudless day, with the sort of crystalline visibility one expects in late September. In his little book of 1922, *The High Peaks of the Adirondacks*, Robert Marshall rated the view from Panther as the tenth best in the High Peaks: "Being so near [Santanoni], the view is of necessity much like it. However, Santanoni cuts off many of those ponds which add so much to the view from the higher summit. The passes do not stand out so well from Panther either. However you can see three peaks from Panther which are invisible from Santanoni, and you get a better view of the upper part of Cold River."[5]

In 1966 Panther offered a spectacular view, but not one entirely unobstructed by the gnarly firs: Sewards to the northwest, the highest peaks to the east and not easy to see. The best view was due south, toward nearby Santanoni, and to the southwest, directly over Cooch, beyond which we could see the north end of Long Lake and Kempshall. It was jaw dropping, it was deeply, emotionally compelling. By the time Robert Marshall stood on the summit of Panther, he had enjoyed the views from many other Adirondack peaks. Hence, I suspect, his rather mechanical account of the view. For me, a rank newbie, it was soul stirring and unforgettable. As with many others before and after me, my first time on a High Peak with an at least partially unobstructed view was the one that grabbed me by the throat and never let go.

The sense of transcendence, the spiritual jolt, on an Adirondack peak derives from a host of factors. There is the view itself, the seemingly endless succession of forested ridges and bare rock, punctuated by the glint of ponds and lakes. There is the nearly complete absence of discernable human imprint on the landscape. But an essential element is also the work it took to get to the summit. For someone who gets to the summit of Whiteface in a car, the view is infinitely different from that greeting someone who hikes up from, say, Connery Pond, as I did three years later. Add to that the fact that my first two peaks involved bushwhacking, the likes of which I had never previously experienced.

To be sure, it wasn't the north face of the Eiger. It wasn't Grand Teton. It wasn't even Katahdin. It was a pair of rounded, northeastern American peaks mostly covered with trees. No technical climbing, specialized skills, or fancy hardware required. In almost all circumstances, in the summer anyway, no element of significant danger; if you got lost, the worst thing that could happen was probably an uncomfortable night in the woods without your sleeping bag. All you needed was a pair of sturdy boots, long sleeves and pants, reasonable stamina, map-and-compass capability, and plenty of time. But the impression it left—not to dwell on the sense of accomplishment—remains indelibly powerful.

After an hour or so on the summit, we started a long, slow, and laborious descent, heading west of north, angling directly back toward Cold River Flow, bypassing Cooch. At first, it was just like the ascent, only going downhill: barking our shins on fallen trees we couldn't see, fighting our way through the thick spruces, often balancing two or three feet off the ground, which we also couldn't see. Eventually, we slogged our way to below the blowdown line and moved faster, finally rejoining the rest of our group just after sunset.

Having heard of the Forty-Sixers only a few weeks earlier, I was on my way. This sort of checklist is almost inevitably appealing to the adolescent mind (and to many others apparently).[6] Couchsachraga was my first and Panther my second. Later that summer, my good fortune continued, and I joined Don McMullen on another peak-bagging trip, this time to the Great Range. It was the classic traverse, starting at the Garden and heading for Lower Wolfjaw, followed by Upper Wolfjaw, Armstrong, and Gothics, with a descent from Gothics after dark to the Orebed Brook lean-to, now long since removed by the State.

The lean-to was already occupied by a handful of Boy Scouts, and there was no water, except what we were carrying and what the Scouts generously offered to share. We fixed dinner with what we and they had on hand and crowded into the lean-to, where the Scouts, continuing with the spirit of accommodation often found in the backcountry, made room for us. First thing the next morning, one of the Scouts and I cleared out our packs and loaded them with empty canteens. We descended the Orebed Brook trail to the first water, filled the canteens, and humped back to the lean-to. In those days, we had yet to hear of giardia, and we drank any running water we came across. I still do.

In the next two days, we climbed over Saddleback, Basin, Haystack, and Marcy. Then down the Feldspar Brook trail to the Opalescent and Lake Colden and Flowed Lands. No rain since leaving the Garden, but the first night at the Calamity lean-to, we were hit with a midnight cloudburst, which made the hike out on the Calamity Brook trail to Upper Works, always muddy, even sloppier than usual. I think I had my eighteenth birthday on that trip.

In 1966, the High Peaks were not as crowded as they are today. We didn't have the summit of Marcy all to ourselves, but there were few other hikers there. I think we were the only people on Haystack. Ditto for Basin (which remains one of my favorite vistas). That would be virtually impossible on any dry summer day in the 2020s.

With an understanding that I would be assigned to the older campers, known at AWC as the Outpost, I eagerly signed up for the summer of 1967. I could plan my own trips and get paid to become a Forty-Sixer. Given my age and the location of AWC, it's not surprising that I focused almost exclusively on the High Peaks. Encouraged to look elsewhere for my wilderness by Bill Verner, who began gently suggesting that wilderness was more than good views, I also took campers to the Siamese Ponds and the Five Ponds, and a couple of times down the Northville-Placid Trail from Long Lake as far as Piseco. The camp would drop me and my campers off at any trailhead and meet us wherever we planned to finish; for most trips, we would start or finish at Plumley's, trying to minimize the car time. I climbed my forty-sixth peak, Esther, in June 1969.[7]

My time at AWC occupied all the summers around my college years, plus one—1966 through 1971. The camp folded after my last summer there. The expenses of running a large camp on an isolated lakeside about nine miles from the nearest town and public road were simply too much. When every bite of food and every bit of supplies has to be transported by boat, with the exception of nonperishable stuff that could be trucked down over the ice in the winter, the costs were astronomical, and after 1971 the camp owners surrendered to the reality of the bottom line. The property is now in the Forest Preserve.

During the late 1960s, when I was a counselor at AWC, and the early '70s, when I was working at the Adirondack Museum, and into the '80s and beyond, after I was able to buy land on the east shore of Long Lake, I camped as often as I could manage in the two ranges on either side of Cold River, the Sewards and Santanonis, always approaching from the Long Lake side, often solo. This country was relatively underused then. I never encountered another person on the summit of a Cold River Peak.

I bushwhacked up these mountains from every direction, in small groups or alone. I have often wondered if I might be the only hiker (or at least the first) to climb each of the Sewards individually, that is, one peak on one day without going to the other nearby peaks. I climbed Seward from the height of Ouluska Pass and from Calkins Creek, Donaldson from the west via Boulder Brook, and Emmons straight from Seward lean-to on Cold River and from Ouluska Pass Brook. I have also climbed Emmons and Donaldson without going to Seward, and Seward and Donaldson without bagging Emmons.

The recommended route for the three western Sewards was to start on the Ward Brook fire road and climb Seward from the north, then along the ridge to Donaldson and Emmons. There was a herd path following this route. Most people then reversed and went back to Ward Brook the same way they had ascended. Another option once you were on Emmons was to hike back to the height of land between Emmons and Donaldson and drop down the east side of the ridge, eventually reaching a substantial old lumber clearing on the east side of Ouluska Pass Brook. Bill Verner named this clearing the World's Largest Raspberry Patch, and that's what we always called it.

The Sewards were logged by the Santa Clara Lumber Company from around 1890 to 1920. This clearing was an artifact of that era. Scattered close to the surface were broken bottles, horseshoes, lanterns, the occasional sole of a boot, and all sorts of other debris. Santa Clara, cutting mostly softwoods for pulp, logged up to the very top of the ridge lines. They loaded logs onto sleds which were lowered on ice roads, perfectly straight, using Barienger brakes.[8] In the '60s and for many years later, these roads were still traceable below the blowdown, and were a handy way up to and down from the blowdown line. I once stumbled across the steel cylinders of a rusty Barienger brake high on the east side of Donaldson. I saw them again a summer or two later. Given their weight, I expect they are still there.

By the time I wrapped up my fifth year at AWC, I considered myself an environmentalist and a wilderness obsessive. I used the word wilderness with casual abandon, and in my mind it was especially associated with the Cold River peaks. As far as I was concerned, wilderness was a reality, both historically and biologically. If I was deep into the west slope of the Sewards or in the remote Callahan Brook bowl west of Santanoni, far from the nearest paved road, I was in the wilderness. The evidence of a massive industrial intrusion into those forests—so obviously manifest in the Santa Clara tote roads and clearings and the debris lying on the ground, or close to the surface, at the Raspberry Patch—did not suggest to me that my environmental values were inconsistent or illogical. My friends and I talked about the "wilderness experience" and bitterly condemned the State's use of heavy machinery in the late '60s to build a horse trail along the south side of Cold River as a violation of that experience. That the Bombardier mini-bulldozer I once came across there was a gas-sucking, inappropriate intrusion into the otherwise remote and peaceful Cold River Canyon seems incontestable, but its relationship to the heavily capitalized logging operations that had worked the same spot only a few decades earlier eluded me. (I resisted the urge to dump sugar into that machine's gas tank.) So did the irony involved in the fact that the Cold River horse-trail system mapped out by the Conservation Department was using some of the same tote roads I found so useful in peak bagging.

Wilderness was less a place than a feeling, a belief. I had read little of the primary or secondary literature on American wilderness. I had never heard the term "cultural construction." But I knew that sitting on a log across the upper reaches of Ouluska Pass Brook, sipping a touch of bourbon with my hiking buddy Bob Pettee, profoundly aware of the peaks around us, the clean water running past our boots, and the distance to any town or road, I was in a place that moved me spiritually. We weren't in a place remotely comparable to Robert Marshall's Alaska (about which more to come), but it was good enough for us. With wilderness, everything is relative. We are all looking for meaning in our lives, however illogically or vaguely we may define that word and however haphazardly or unconsciously we conduct the search. My five summers at AWC had led me to a sense of meaning I found nowhere else. That sense has evolved over the decades, but it began with the trek to Panther in 1966, a week or so before my eighteenth birthday. And it grew significantly when I began work as Bill Verner's Research Assistant at the Adirondack Museum.

5

Adirondack Museum

My last season as a counselor at AWC was in 1971. I had recently dropped out of graduate school (University of California at Davis, where I had pursued, unenthusiastically, a PhD in English for less than one term before slipping anonymously away from Davis the day before December exams began and pointing my tinny Datsun pickup back toward the East). This was followed by a few weeks stocking shelves in a college bookstore in Iowa and then pumping gas in western Massachusetts before the final summer at AWC.

I was wondering what to do with myself. On a whim, I approached Bill Verner and asked if there might be a job for me at the Adirondack Museum. It was another of those random decisions, making sense at the time—a hasty query in response to a pressing need to find a job—that changed my life. It was luck and coincidence that took me to the Adirondacks in 1966, and it was the same combination that led me to the Museum in the fall of 1971. Bill needed help in planning what was eventually the Woods and Waters exhibit (still there, with minimal changes having been made over the nearly fifty years since it opened in the late '70s), and Jerry Swinney had the budget to hire me. It didn't take much: I was paid $100 per week. But you could live on that in Long Lake, where Bill and Abbie Verner lived with their two daughters.

With Bill supplying lengthy reading lists, I began to bone up on the history of the Adirondacks. My assignment was to research and then write a report on how people had interacted with Adirondack wildlife, beginning as far back as we could find any documentation and running up to the early twentieth century. Given what animals were even mentioned in the primary documents—exploration and travel narratives, official reports, sporting narratives, and more— this inevitably meant large mammals; few people then paid much attention to mice, bats, squirrels, and the like. The substance of this report, after further research and much rewriting, became a book, *Wildlife and Wilderness: A History of Adirondack Mammals* that I published in 1993.[1]

What's important here is how critical this research was to my subsequent career, my evolving thinking about wilderness, and even to my eventual participation in *Protect*. Bill's command of the literature was prodigious, and the Museum possessed a comprehensive and constantly growing collection of Adirondackiana. The librarian, Marcia Smith, let me set up a table in the midst of her domain. I could take anything off the shelf, read it, take notes, and move on. I read everything Bill could think of that involved a record of anyone shooting, chasing, or even thinking about mammals in the Adirondacks. The reading snowballed, one item leading to another. For the most part, I was consuming countless accounts of hunting white-tailed deer and getting familiar with the huge and wonderful canon of nineteenth-century Adirondack literature.

After dipping into the handful of Colonial-era documents that shed some (limited) light on how Native Americans might have interacted with the wildlife of northern New York and on what was going on outside the Adirondacks or on the fringe before the early nineteenth century, I dived into a mass of utterly fascinating accounts of travel and field sports in the Adirondacks written at a time when the region barely showed up on New York maps. In reports to the New York legislature, geologist Ebenezer Emmons described explorations in the central Adirondacks that announced the existence of this huge, virtually unknown territory. This inspired the classic mid-century travel narratives by Charles Fenno Hoffman, Joel T. Headley, Samuel Hammond, Alfred Street, Thomas Bangs Thorpe, William H. H. Murray, and many others.[2] They were exhilarating.

On index cards, I wrote down whatever they had to say about wildlife: countless descriptions of jacklighting, hounding, and still-hunting deer, to be sure, but also encounters with black bears, wolves, cougars (almost always called panthers or mountain lions), beavers, otters, and moose, with the occasional mention of other furbearers like minks or fishers. My primary charge was to record what nineteenth-century sportsmen thought about these mammals. But even more interesting to me throughout this project were the accounts of the adventures of "sports," as they were usually called, people who were rowed and led by local guides into remote spots known only to a handful of trappers and stubborn farmers settled in and around the nascent villages of Long Lake, Newcomb, Old Forge, Keene, or Saranac Lake. They climbed Marcy and had the summit entirely to themselves. Mostly they sat in guide boats and were rowed by famous guides like Mitchell Sabattis and Harvey Moody to remote spots like the headwaters of the Bog River, where they found not a trace of any previous human presence.

I read the entire Colvin canon, of which the Museum's library had a complete run: book-length (often a rather large book!) annual reports as well as some of only a few pages. Verplanck Colvin is perhaps the most fascinating of a large stable of nineteenth-century Adirondack writers. I am but one of many (including Robert Marshall, as explained in a subsequent chapter) who have enviously read his reports about perilous bushwhacks in the High Peaks and wintry explorations of frozen lakes in the western Adirondacks. Beginning in the early 1870s and running through the first few years of the twentieth century, Colvin produced a stunning body of work. As with the authors of the sporting narratives, I was ostensibly searching for whatever happened whenever he or his guides came upon a deer or a cougar. But for the most part I was drooling over what it was like to stand, for example, on the edge of Lake Tear of the Clouds in 1872 and be quite convinced that no one had ever been there before. Whether Colvin and his companions were actually the first to encounter Lake Tear, given the fact that people had lived in New York for millennia before he entered the scene, is another matter. Perhaps he, like so many others before and after, was projecting an obsession with exploration of putatively virgin land onto his experience.[3]

Prodded by Bill Verner, I was learning to see the value of wilderness environments outside the seductive scenery of the High Peaks, places not defined by peak bagging and the distant vista: wetlands and dense forests, for example, where the intricate details of natural process seemed uncontaminated by human interference. Reading the classic narratives of the nineteenth century, I also came to think, among other things, of wilderness as a point of entry into historical understanding. This is what it was like 200 years ago, I thought, and what I'm doing when I'm deep in the woods might reflect that experience. It was not a particularly sophisticated or well-considered understanding of wilderness. It was performative. I was not thinking about inconsistencies, about class privilege or about the nearly completely erased Indigenous presence. This would come later.

6

Roderick Nash

At the Museum, I worked on my wildlife report and read and reread the thrilling narratives of Headley, Colvin, and dozens of other Adirondack writers. Bill Verner and I discussed the recently approved Adirondack Park State Land Master Plan (1972) with its creation of designated Wilderness Areas in the Forest Preserve. And the book that both of us used to frame much of our thinking about both wilderness history and wilderness policy was Roderick Nash's *Wilderness and the American Mind*. With its prodigious catalog of primary sources, its clarity of argument and purpose, and its energy, not to mention the fact that Nash paid particular attention to the Adirondacks and to our own Robert Marshall, it grabbed us powerfully.

Nash's path-breaking book begins with certain key assumptions: wilderness is real, it's a treasure, and it has a history inextricably bound with America's. Nash shows how Euro-American culture began with a fear of wilderness and gradually moved to appreciation. The triumphal arc of that interpretation of American history utterly seduced me. His reading of the American encounter with wilderness and how values changed over time runs from the early seventeenth century through the late twentieth, and it follows a neat and satisfying narrative curve. Too neat and satisfying, many readers later concluded.

On a chilly November day in 1620, as the Pilgrims first cast their eyes on the shores of Massachusetts, they saw only, in William Bradford's recollection of that historic moment, "a hideous and desolate wilderness," something they wanted to eliminate as quickly as possible. Bradford describes the place only in terms of what he perceives it not to be: he and his fellow Pilgrims "now had no friends to welcome them nor inns to entertain or refresh their weatherbeaten bodies: no houses or much less town to repair to, to seek for succour."[1] Nash understands Bradford's antipathy to what he sees and his longing for what he doesn't see as the onset of a "tradition of repugnance."[2] The easily demonstrable fact that human societies enjoying a rich culture stretching back over centuries already lived on Cape Cod, where the Pilgrims first landed, did not occur to Bradford. One reason it did not is that eastern Massachusetts's Indigenous populations had been horribly reduced over the previous decade or so by diseases communicated to them by European sailors and commercial fishermen. Historian Francis Jennings neatly distills this tragic irony: "Europeans did not find a wilderness here; rather, however involuntarily, they made one."[3]

It's true, of course, that these Englishmen and -women were in need of shelter, and the Massachusetts seacoast in late November was hardly a salubrious spot for such unprepared, ill-supplied folk. They needed food and shelter, and they needed these immediately; their situation was deadly dangerous, and many of them died before they could see the following spring. But it's also the case that the understandings of Bradford and his companions were predetermined by an already prevalent mythology and the Protestant certainty that the material world is both deceptive and seductive. Bradford certainly did not invent the pattern he invoked, but he gave it an eloquent and telling expression, establishing it as an archetypal element in our early literature.[4]

As the English settlers slowly bent what was to them a New World to be agriculturally productive and as the Indigenous cultures either retreated or were wiped out by disease and violence, attitudes began to evolve. The arrival of the eighteenth-century fascination with the workings of nature convinced some of the more literate Americans that the untamed wilderness over the next ridge might be not threatening but interesting and engaging. The naturalist John Bartram, for example, typified a new way of responding to the wilderness, seeing it as a place for research and even as a reflection of the magnificence of a benign and omnipotent creator.[5] His son, William, represents the watershed between the Enlightenment and proto-Romantic values, taking his father's nascent appreciation several steps further, reveling in the lushness and complexity of the Florida backcountry and southern Appalachia.[6]

William Bartram's *Travels* (1791) is one of the first and certainly one of the finest of early American wilderness narratives. It has its share of ambivalence, of course, but its warm embrace of the magnificent diversity of wild nature, even when it is threatening or even dangerous—as with his famous encounter with alligators on Florida's St. John River—offers a striking contrast to the hostility and fear displayed only a hundred years earlier.[7] It is thus an important point in the linear and triumphal narrative of evolving attitudes laid out by Nash.

The growing American appreciation of the splendors of their recently independent nation, which was inspired by intellectual currents radiating from Europe, was slow to penetrate the wider culture. As Alexis de Tocqueville noted in the early nineteenth century,

> In Europe people talk a great deal about the wilds of America, but the Americans themselves never think about them: they are insensible to the wonders of inanimate nature and they may be said not to perceive the mighty forests that surround them till they fall beneath the hatchet. Their eyes are fixed upon another sight . . . draining the swamps, turning the course of rivers, peopling solitudes, and subduing nature.[8]

The vast majority of Americans continued to see the natural world as a locus of livelihood if not the source of wealth. They did not fear the wilderness as Bradford and the Pilgrims had, but they mostly saw it as something to be cultivated, logged, mined, or otherwise harnessed or eliminated to further their needs.

But as every account of American literature and art reminds us, the germs of European thinking slowly penetrated the literate American classes. From the mountains of New Hampshire to the Hudson Valley and the springs of Virginia, some Americans began to find spiritual solace in the wilderness. And a major destination for those Americans encouraged by the spirit of the day to redeem body and soul in the wilderness was the part of northern New York described by Ebenezer Emmons in the late 1830s and early '40s. They channeled a Words-worthian anxiety about the frenetic pace of American commerce and industry and found physical, moral, and spiritual redemption in the apparently untouched corners of the eastern wilderness.

In the 1840s, reflecting the growing authority of romantic values among middle-class Americans, Joel T. Headley responded thus to the unsettled wilderness of the Adirondacks:

> I love the freedom of the wilderness and the absence of conventional forms there. I love the long stretch through the forest on foot, and the thrilling, glorious prospect from some hoary mountain top. I love it, and know it better for me than the thronged city, aye better for soul and body both. . . . I believe every man degenerates without frequent communion with nature. It is one of the open books of God, and even more replete with instructions than anything ever penned by man. A single tree standing alone, and waving all day long its green crown in the summer wind, is to me fuller of meaning and instruction than the crowded mart or gorgeously built town.[9]

This passage is almost a catalog of popular romantic tenets, even clichés, with its anti-urbanism, its faith in the divinity of nature and its power to shield mind, body, and soul from the miseries of mo-dernity. But while its rhetoric and its sentiments may be unoriginal, it is an honest and compelling expression of something many of us feel today. It's worth noting what led Headley to the wilderness in the first place: working hard in the frenetic carnival of Jacksonian America, he suffered what he called an "attack on the brain," what we would probably identify as a nervous breakdown.[10] His expeditions into the heart of the Adirondacks did the trick, and he went on to a prodigiously productive, healthy, and long life.[11] Headley is a pivotal touchstone in Nash's account of how the romantic inclination to see value in wild nature penetrated America's eastern establishment in the mid-nineteenth century.[12]

Nash likewise invokes the Hudson River canvases of Thomas Cole, Asher B. Durand, and their disciples to illustrate the growing

American fascination with the wilderness and the association of the wilderness with a formative national identity.[13] Americans, especially those who traveled to Europe, were culturally insecure, all too aware that their new country lacked classical antiquities like the Parthenon or splendid cathedrals like Notre Dame, and feared that this absence might mean that a great American literature and artistic tradition would never develop. The obvious substitute was the grandeur of the American land itself. Defenders of American culture routinely offered the sublimity of the wilderness as the subject of a uniquely American artistic regime.[14]

Headley, Cole, and others of their era set the stage for the achievements—both political and literary—of John Muir, an iconic player in *Wilderness and the American Mind* and its take on American history. For many years, Muir explored the high country of the Sierras and wrote about his adventures in a prose style that was intensely personal, engaging, and evocative. He took Edmund Burke's rhetoric of the sublime and beautiful, a powerful force in our literature for at least a century, combined it with Ruskinian metaphors of the Gothic, and pushed them to perhaps their most penetrating application. He also adapted the New Testament rhetoric of salvation, relentlessly forced on him by his Calvinist father, and used it to recount the personal sense of redemption he experienced in places like the jaw-droppingly magnificent Yosemite Valley. Both of these literary traditions—of the Burkean sublime and of Christian redemption—were already well established in American wilderness literature. But Muir was the master of both, achieving a mass appeal and invoking a warmth and originality that seldom lapsed into the overwrought prose of many of his contemporaries.[15]

As Nash and others have recounted, Muir was the first to translate his love of the American wilderness into a political force. The story of the battle over the Hetch Hetchy Valley, a part of the recently established Yosemite National Park, is an often-repeated episode in the American wilderness narrative, and it gets a full chapter in *Wilderness and the American Mind*.[16] John Muir organized a nationwide campaign in an effort to prevent the construction of a huge dam and reservoir in one of America's first national parks. It was a struggle that lasted for years. In rallying the troops, Muir and his comrades invoked a litany of arguments that have become almost routine parts of the wilderness defense. The critical point was that with its scenic magnificence and unspoiled purity, the valley was a temple, a sacred grove for a new American religion. Turning even a part of it into a manmade reservoir was a sacrilege.

Muir and his allies lost the battle, with final Congressional approval of the dam coming in 1913. But the fierceness with which the valley's advocates labored in its defense are convincingly invoked by Nash as solid evidence of the extent to which the idea of wilderness had become a force to reckon with in American political life. The reality of a national wilderness constituency begins with Hetch Hetchy. And in the Roderick Nash version of growing wilderness advocacy, Hetch Hetchy marks the point where (with some important exceptions) he turns from individual narratives of appreciation, like Headley's in the mid-nineteenth century or Muir's a few decades later, to the stage of politics and the campaign waged by Muir's spiritual descendants to extend statutory, reliable, and permanent (though nothing is truly permanent, of course) protections to what remained of the American wilderness.

One of the first elements in this process was the establishment of the National Park Service in 1916, reflecting the resolve of conservationists that a desecration like the dam at Hetch Hetchy would never happen again. The origin of the Parks themselves goes back to 1872, when Yellowstone National Park became the first such protected space in the world. At that point the nation was far from sure what this and subsequently established parks were for. Were the parks created so that western railroads could sell tickets to exotic destinations? Or was their purpose, in fact, the protection of untamed nature for the benefit of all Americans?[17] It took a long time for the country or the Park Service to make up their minds, but by the 1930s, a growing constituency demanded that neither large projects like dams nor opulent amenities for tourists should further compromise the integrity of the Parks. Eventually, this sentiment prevailed, and the National Parks, appreciated and visited by millions of Americans every year, now protect a significant portion of the remaining American wilderness.

A critical new feature of how Americans thought about their lands and what should be done with them derived from the nascent science of ecology. Nineteenth-century enthusiasts like Headley found the appeal of wilderness in scenery, in the pantheistic conviction that god dwelt in the land, in the rugged allure of hunting and fishing, or in the opportunity to get away from the raucous din of urban America. With the emergence of ecology as a science and a way of thinking, beginning roughly in the 1920s, the idea of nature as habitat, as a complex interconnected web of relationships like predation, energy exchange, and evolution, expanded the reasons for valuing and protecting wild nature.[18]

The paradigmatic figure in Nash's account of this development is Aldo Leopold, whose research in wildlife management and the subtle nuances of predator-prey relationships, among many other things, suggested that nature is a lot more complex than most people had previously grasped. If we are to understand nature, to get closer to the innermost workings of its cycles and intricate patterns, argued Leopold, then we had better keep some of our natural heritage in an uncorrupted condition. Shortly after World War I, Leopold, then working for the US Forest Service, began advocating within the federal bureaucracy for setting part of the vast public domain aside as undeveloped wilderness.[19]

Wilderness as scenery or as locus for recreation or as artifact of American history did not disappear as reasons for protection, but the gradual penetration of ecological thinking into the wider culture added a powerful new argument to the array of justifications earlier employed by Muir and his allies when they advocated for the sanctity of Hetch Hetchy. It was the emergence of ecological thinking that expanded the National Park idea beyond conventionally picturesque places like Yellowstone and Yosemite to embrace hitherto ignored or even disdained places like Florida's Everglades, which became a National Park in 1947.[20] A couple of generations earlier, when the National Park idea was first germinating, any proposal to protect what most people considered a huge swamp would have gained minimal traction, but with the Everglades, the nation committed to protecting wetlands and wildlife habitats, *as such.*

Most of Leopold's focus was on the National Forests. Since they had been established almost exclusively as sites for extractive industries like logging and mining, the idea of setting some of their acreage aside as wilderness was radical thinking.[21] It did not immediately become policy, of course, but Leopold found a few allies in the Forest Service, including the Adirondacks' Robert Marshall. Meanwhile, Leopold, whose original impulse had been to protect habitat and terrain for backcountry hunting and field sports, began to ponder the whole, complex picture of what wilderness meant to him, why he should dedicate himself to protecting it, and what a preservation strategy might entail. He concluded that this was an issue of public welfare, that Americans would all be better off, that their lives would benefit once wilderness was protected.[22]

The difficulty was in convincing Congress that it was to their advantage to slow down the process of ruthless exploitation that had defined the national project for so many centuries and that had provided piles of cold cash for Washington lobbyists. But in a dramatic triumph, wilderness advocates eventually prevailed. The outcome of all this, the twentieth-century consummation of centuries of attitudes that morphed from hostility to love and the climactic moment of Nash's 1967 bestseller, was the National Wilderness Act, signed by President Lyndon Johnson in 1964.[23]

The idea that American attitudes toward wilderness underwent an almost Darwinian evolution from hostility to nuanced appreciation, from William Bradford's horrified recoil to Robert Marshall's demand for protection, culminating in the passage of the Wilderness Act, is the fundamental message of Nash's *Wilderness and the American Mind.* First published over half a century ago and still very much in print in a fourth and much-expanded edition, it is one of the founding books of American environmental history and one of the bestselling books in the history of Yale University Press. Its importance in both the environmental movement itself and in the formative years of the growing subdiscipline of environmental history cannot be overstated. The Yale University Press website hints at the significance of this scholarly landmark: the Los Angeles *Times* identified it as one of the one hundred most influential books of its quarter century; no less an environmentalist than Dave Foreman, one of America's most passionate defenders of wilderness, declared it "a must-read for anyone who wants to understand wilderness and the American conservation movement"; and *Outside Magazine* listed it among the "books that changed our world."[24] Despite the efforts of revisionist historians to illuminate the narrowness of Nash's understanding of cultural history, *Wilderness and the American Mind* continues to exercise enormous power in the American popular imagination.[25]

7

Robert Marshall

One of the delights of my job at the Adirondack Museum was the opportunity to browse the shelves in the library and grab any book that caught my attention, often enough one that had little to do with my wildlife project. One of these was Robert Marshall's *Alaska Wilderness*, an account of his explorations in the remote Brooks Range in northern Alaska between 1929 and 1939. As all Adirondack peak baggers know (or should know), Marshall, his brother George, and Herbert Clark, a Saranac Laker who was both an employee and friend of the Marshall family, were the first recorded hikers to climb what later became known as the Forty-Six High Peaks. What got the Marshall brothers started on their quest to ascend the Adirondack High Peaks was their immersion in the Colvin reports, of which their father had a complete run, conveniently shelved and available at Knollwood, their family retreat on Lower Saranac Lake. Just as I did a half century later, the Marshall brothers devoured the Colvin reports and their often-overhyped accounts of exploring the High Peaks.[1] Then they worked out a rather eccentric scheme to identify and hike to all the summits with an elevation of 4,000 feet or higher. They took a few summers to get them all climbed and had great fun doing it. In 1922 Robert published a brief account of their hikes, *The High Peaks of the Adirondacks*, the first of his many publications.[2]

Marshall went on to a career with the US Forest Service, among other places, and at the time of his tragically early death in 1939 was one of the most productive and dedicated American conservationists of the first half of the twentieth century. He was also one of the first Americans to theorize the importance of wilderness, especially in his article, "The Problem of the Wilderness," published in *Scientific Monthly* in 1930, and was one of three founders of the Wilderness Society, in 1937.[3] When it comes to understanding the American love affair with wilderness, including Adirondack wilderness, Marshall is a figure to be reckoned with.

Because Marshall's environmentalist roots were in the Adirondacks, the Museum's library had copies of each of his two books about his time in Alaska.[4] I devoured *Alaska Wilderness* (the other book was *Arctic Village*): its accounts of hiking, climbing, and exploring in a region Marshall thought he was the first White man ever to lay eyes on thrilled me. I also liked the way he slipped in the occasional reference to the Adirondacks; at one point, searching for a way to illustrate the notion of "natural scenery of surpassing beauty," he picked Yosemite, Glacier National Park, the Grand Canyon, and our Avalanche Pass as prime examples.[5]

Alaska Wilderness, when read along with "The Problem of Wilderness," provides a window into his thinking. Marshall used the word wilderness in an unreflective way. He had no doubt that such a thing existed, that it was a useful and unproblematic term, that wilderness referred to a specific kind of place, and that even though there might be debates over what did or did not qualify as wilderness, he knew it when he saw it. When he wrote of "The Problem of the Wilderness," the problem was getting enough of it adequately protected, not the possibility, as subsequent scholarship has suggested, that the entire notion of wilderness was culturally or socially constructed or that the inclination to treasure it and secure its preservation reflected an agenda at least partially originating in ethnocentric or elitist priorities.

Throughout his published and unpublished writing, Marshall frequently returned to the notion that time in the wilderness evokes a connection with America's frontier past, especially the epic explorations of Lewis and Clark. James Glover, Marshall's first biographer, quotes George Marshall, who recalled in an interview that the two Marshall boys designed a "whole series of imaginative games, some involving sports, some involving a kind of Lewis and Clark expedition."[6]

In an essay originally composed as part of the public relations materials for *Arctic Village* and subsequently adapted by George Marshall as the opening chapter for the posthumously published *Alaska Wilderness*, Robert Marshall signaled the importance of Lewis and Clark to him and mentioned a chief source of his fascination with their explorations:

> The story logically begins in New York when I was eleven years old and in bed with pneumonia. In order to keep me pacified they read me a story by one Captain Ralph Bonehill entitled *Pioneer Boys of the Great Northwest*. Thereafter I reread *Pioneer Boys* from one to three times every year for the next ten years. It was a splendid narrative of two lads and their fathers who accidentally joined the Lewis and Clark Expedition and went through all the glorious adventures of the most thrilling of all American explorations. My ideology was definitely formed on the Lewis and Clark pattern, and for quite a period I really felt that while life might still be rather pleasant it could never be the great affair it might have been if I had only been born in time to join the Lewis and Clark expedition.[7]

Ralph Bonehill was one of the many pseudonyms of Edward Stratemeyer, the brains behind a vast and influential publishing empire. Stratemeyer dreamed up the Rover Boys series and was eventually responsible for the Hardy Boys, the Bobbsey Twins, Tom Swift, and Nancy Drew.[8] Stratemeyer's own ideology reflects many of the concerns of muscular Christianity and similar turn-of-the-century movements fearful of a decline in American vigor and virility, especially among the children of the wealthy. As a major figure in the children's literature of the day, Stratemeyer understood his responsibility to be one of instilling a commitment to ruggedness and the active life. "I have no toleration for that which is namby-pamby or wishy-washy in juvenile literature. This is a strenuous age. . . . The best an author can do is to give [readers] a fair proportion of legitimate excitement, and with this a judicious dose of pleasantly prepared information."[9]

The particular book that so inspired the young Robert Marshall, *Pioneer Boys of the Great Northwest, Or With Lewis and Clark across the*

Rockies (first published in 1904), is pretty much what one would expect: a predictable yarn of adventure and exploration, with set-piece descriptions of wild country (here's a response to the Great Falls of the Missouri: "'What grand scenery!' murmured Oscar, as he and Fred gazed around at the falls, the rocky hills, and the great mountains completely covered with snow. 'Yes, and what an immense country it is!' answered his chum. 'I never dreamed it could be so big and so—so sublime.' 'Sublime is just the word, Fred. How little one feels when he looks at these great rocks and those mountains!'").[10] We get exciting adventures on river rapids and perilous mountains, much hunting and fishing, racist accounts of violent encounters and captivity with "marauding savages" routinely portrayed as "thieving," "treacherous," "dirty, foul-smelling," "rascals," and a cheerfully ahistorical account of events.[11] It's easy to see how appealing such a book might have been to a preadolescent White boy from an affluent family. The tale is blithely absurd in its illogicalities and impossible feats of woodcraft.

What's important concerning this book and what it meant to Marshall—to the extent that he identified the fantasies it stimulated as his "ideology"—is how it dwells on the wilderness experience as an *adventure*, one combining spectacular scenery and a window into a comfortably perceived and Eurocentric reading of American history. This is of course an entirely ordinary characteristic of the genre of which it is an example, but the emphasis Marshall himself places on it as a formative influence on his thinking and the fact that his brother George chose to lead off *Alaska Wilderness* with Robert's recollection of it are telling. That the wilderness was an arena for epic adventure and historical understanding constituted a major reason he wanted to protect wilderness throughout the country. For Marshall the wilderness was a place primarily for the experience of an imagined, that is to say, constructed, immersion in a particular narrative of American history, all occurring amid the undeniable scenic grandeur of an ostensibly pristine American landscape. This is the emphasis in *Alaska Wilderness*.

In *Alaska Wilderness*, Marshall explicitly grapples with what we might call his true "Problem of the Wilderness," his awareness that, up to the time before his first trip to Alaska, his adventures—in the

Adirondacks and the Rockies—were not "real" in the sense that he thought Lewis and Clark's or Verplanck Colvin's were: he wasn't the first one there. His solution had been to "rationalize," to convince himself that "mental adventure and physical adventure were in reality the same thing." But, he goes on to say, this stab at self-persuasion was not entirely successful, and he "determined to take a two-month fling at *real* [italics added by me] exploration by spending the major share of the summer of 1929 in what seemed on the map to be the most unknown section of Alaska. . . . I pretended to myself that the real reason for this expedition was to add to the scientific knowledge of tree growth at northern timber line."[12]

During his explorations, he in fact made a point of recording temperatures of both air and soil and pursuing other research into the arboreal ecology of central Alaska. With uncompromising honesty and characteristic self-deprecation, though, Marshall admits that this was a sham and that his research had not led to publication. So, wistfully, he admits that the "real" excuses for wilderness exploration no longer obtain and that one must simply accept that assuring one's "personal happiness" is the only available twentieth-century justification for exploring unmapped territory.[13] (Nonetheless, the narratives stitched together to form *Alaska Wilderness* make much of his mapping and measuring, securing and recording data for everything from elevations to temperatures and the diameters of stunted spruces.)

The dream of being with Lewis and Clark or with Verplanck Colvin in the post–Civil War Adirondacks exercised continuing and compelling authority in Marshall's wilderness fantasies. The importance of Colvin to these fantasies surfaces tellingly in the manuscript of Marshall's unpublished novel, "Island in Oblivion," which now resides in the huge collection of his papers at the Bancroft Library of the University of California. He composed this novel while spending the winter of 1930–31 in Wiseman, Alaska.[14] The novel begins with a young boy growing up in the Adirondacks, early enough for him to have been present as a child at John Brown's 1859 burial at North Elba, and eventually accompanying a Colvin survey team on an expedition up the remote Bog River. The boy decides to become a surveyor like

Colvin and to dedicate his life to wilderness exploration. After the Civil War he finds himself in the Dakota Territory and then sets out to Alaska to prospect for gold. He spends most of his life in Alaska.[15]

That this attempt at a novel is set in an era a couple of generations before his own illustrates Marshall's fixation with the idea of the authentic wilderness experience as something lost, something hopelessly unavailable to a young man of the early twentieth century. Deeply aware of the apparent differences between his explorations and the wilderness of his imagination, he gnawed at his sense that his travels lacked authenticity: planning a trip into the Arctic Divide, he lamented, "Four weeks are not very long for an expedition compared with the two and one half years of the Lewis and Clark Expedition."[16]

Marshall relentlessly pursues the notion that in Alaska he was exploring country "seldom if ever seen by man." Thinking of sites of scenic splendor in the lower forty-eight, he recalls how he had "wished selfishly enough that I might have had the joy of being the first person to discover it," longs for experiences "comparable to those of Lewis and Clark," and asserts that, finally, in Alaska this is what he found.[17] It may be that what he meant was that no *White* man had seen these parts of Alaska; in "The Problem of the Wilderness," he maintains a studied ambivalence about whether Indigenous occupancy or use of a place does or does not disqualify it as wilderness.[18]

In any case, he often, perhaps without thinking, invokes the dream of being first. This fantasy was tellingly questioned by Tishu Ulen, an Innuit woman who was a friend of Marshall's when he lived in Wiseman and who recalled Marshall decades after his death. Speaking of the remote country of the Central Brooks Range, which Marshall described in extensive and loving detail in the articles that eventually became *Alaska Wilderness*, she observed, "'No, he wasn't the first to travel any of the country. I think he just imagined he was.'"[19] Marshall, in fact, constructed his wilderness, and Tishu Ulen recognized this to have been the case.

The document in which Marshall most clearly organized and articulated his thoughts about wilderness was the article he wrote for *Scientific Monthly* in 1930.[20] Here he makes an important transition, from the euphoric adventures he associates with an earlier era to the administrative and bureaucratic realities of a twentieth-century culture in need of a way to relive those adventures. This maneuver anticipated a similar move adopted by Roderick Nash in *Wilderness and the American Mind*. In Marshall's view, the nineteenth century and earlier constituted an era of authentic adventure and wilderness immersion. The twentieth was to be a time of policy and management to ensure the protection of wilderness as "a region which contains no permanent inhabitants, possesses no possibility of conveyance by any mechanical means and is sufficiently spacious that a person in crossing it must have the experience of sleeping out." That is, wilderness is a place where a person *does*. It need not be pristine but should "preserve as nearly as possible the primitive environment."

What about the history of people who live or have lived there, people who are not simply visiting and don't have a primary home somewhere else? Marshall passes over the millennia of pre-European occupation of the vast North American continent, implying that because Indigenous technology was not the same as that of the Europeans their presence did not diminish the wilderness status of the places they lived. He didn't deny that Native people lived in the wilderness; what was important to him was that their occupation of the wilderness did not mean the elimination of opportunities for a wilderness experience for Americans like him. He didn't say it explicitly, but in his understanding Lewis and Clark were experiencing wilderness adventures; Sacajawea, because she didn't live in the White world and did not have a home back East to which she would eventually return, was not. Just as it was for Headley and his generation, Marshall's wilderness was a place one goes to for adventure and redemption but then leaves after reveling in the wilderness experience.

The way Marshall avoids the reality and history of Indigenous people and their lives in what he is determined to see as the era of pre-European purity is a monumental embarrassment to any modern reader, especially when we consider how attuned he was to the realities of antisemitism and racial discrimination in his own era.[21] He acknowledges the Indigenous presence but seems unaware of their modifications of the land: "Wild animals still browsed in unmolested meadows and the forests still grew and moldered and grew again precisely as they had done for undetermined centuries." Marshall did not admit, if he even knew it, that Indigenous Americans skillfully and routinely manipulated the various environments in which they lived: among other things, they altered forests with carefully planned fires, they planted crops, they irrigated arid lands, and they actively promoted some and discouraged other wildlife populations.[22] Although he is aware of the Indian presence, he removes them from the cultural category of the "civilized," by which he means everything he went to the wilderness to get away from. Indigenous people were somehow not living in history.[23]

Marshall's reluctance to ponder the history of Indigenous people living and working in the pre-contact North American landscape brings us to a question that has vexed Adirondack historians for at least a century. And it has clouded debates over whether or not wilderness can even be associated with the Adirondacks. T. Morris Longstreth, the first of a series of amateur historians to make a stab at a regional history, assembled zero documentary evidence and concluded that warfare between New York and Canadian Native groups led both sides to see the Adirondacks as a dangerous battleground and thus not suitable for year-round settlement. It's a reductive, cartoonish version of history: "Villages grew on the outskirts, but, within the confines of the mountains, the frequent massacres prevented all settlements. All was dark and roving and the tomahawk never rested in a truce."[24]

Alfred Donaldson's two-volume *History of the Adirondacks* introduced the idea of seasonality, based on an assumption that Native Americans would have avoided the Adirondacks in the winter. This enabled Donaldson to ignore an obvious conclusion to be drawn from the deadly encounter, in 1609, between Samuel de Champlain, accompanied by Algonquians from the St. Lawrence, and a band of Iroquoians on the western shore of the lake that Champlain named for himself. Both the Algonquian and Iroquoian people in Champlain's narrative clearly knew the region, yet Donaldson does not credit them with "discovering" the Adirondacks. The fact that Donaldson believed them to pass through the Adirondacks only in the summer somehow meant that their relationship with the region was less than real: "The consensus of authoritative opinion seems to be that the Indians never made any part of what is now the Adirondack Park their permanent home."[25]

From the fifteenth century to the present, the notion that Europeans "discovered" lands where non-European people already thrived has been increasingly problematic and is tragically intertwined in the triumphal narrative of steadily growing wilderness appreciation advanced by Roderick Nash and popular to this day. That most Adirondack historians ignored a part in the Adirondack story for Native Americans has now been well established in Melissa Otis's *Rural Indigenousness: A History of Iroquoian and Algonquian Peoples in the Adirondacks.*[26] Otis methodically lays out extensive evidence of several Native cultures occupying the Adirondacks, and she addresses the notion of seasonality. It doesn't take a weather scientist to recognize that the climate—with clear implications for both efficient hunting and agricultural productivity—in the lowlands around the Adirondacks is milder and more comfortable than it is in the central Adirondacks. This obvious fact led White historians to conclude—too quickly, perhaps—that Native peoples must have hunted or traveled through what is now the Park only in the summer or at least not in the winter. This along with the apparent absence of Native groups (except for a few scattered families) in the Adirondacks in the 1830s, when White Americans first began serious exploration, mapping, and settlement, led to the facile assumption that Native cultures had no serious claim on or even knowledge of the region. No one doubted that Native hunters, trappers, and war parties had passed through now and then, but for reasons unexamined by nearly everyone, too many of us unthinkingly concluded that this seasonal use was not enough to establish an authentic presence.[27]

Among the many consequences of this erasure was the appealing deduction on the part of White people that the Adirondacks belonged to no one, that the region was there for the taking. Historian William Cronon addresses this phenomenon in his *Changes in the Land: Indians, Colonists, and the Ecology of New England* and explains the difference between Anglo-American notions of individual property ownership, with our concern for titles and precise boundaries, and the Native concept of "collective sovereignty." Or as Pekka Hämäläinen, a Finnish scholar of Indigenous-European contact in North America, notes, "Land was used rather than owned, allowing a flexible property regime."[28] The Native understanding involved the idea that a tribal group—Mohawks, for example—exercised a territorial claim on a place and that they could try to prevent an alien group—Algonquians from the St. Lawrence, for example—from hunting in or otherwise intruding into that place. This could explain the conflict between Iroquoians and Algonquians described by Champlain. The point is that the Adirondacks were part of a vast territory used and known by Indigenous cultures. They did not write out legalistic titles with precise metes and bounds and assign lots to individual owners, but they understood that certain territory was part of a communal domain, a commons, and was available to one's own group's exploitation. Europeans and, later, Americans quite failed to grasp Native understandings of communal sovereignty, and the Native cultures were baffled by English notions of individual ownership.

Cross-cultural misunderstandings, added to what we now understand to have been massive rates of disease-caused mortality among virtually every Native society in North America following first contact, led too many Anglos—explorers, settlers, historians, and wilderness enthusiasts like Robert Marshall—to write Native cultures out of their narratives of virgin land waiting for either exploitation or preservation. But such a failure to grasp Native cultures and their history and its persistence in recent wilderness debates need not lead us to dismiss any discussion of wilderness and its protection. We must acknowledge Marshall's blindness to historic realities without casting aside his prescription for a modern culture in need of the spiritual satisfaction that wilderness can offer. Marshall was profoundly aware of the spiritual emptiness he perceived all around him in twentieth-century America and was passionately in search of an antidote.

In "The Problem of the Wilderness," Marshall discusses his understandable regret that the dominant forces in American culture had altered so much of the landscape that greeted Europeans and European Americans and devotes most of his essay to the need to protect wilderness for future generations. Given his roots in the Adirondacks, he also, though only implicitly, allows that the wilderness experience he so craved did not depend on virginity, real or imagined. Like him, we can reasonably and logically talk of wilderness in an Adirondack forest that in the past was the domain of Indigenous people and more recently was the locus for extensive, often ruthless logging.

In "The Problem of Wilderness," moreover, Marshall outlines precisely how essential wilderness is to an increasingly urban culture: wilderness recreation leads to improved health, promotes self-sufficiency, and encourages intellectual and spiritual growth. He anticipates most of the arguments against wilderness protection and declares that America must immediately begin to protect remaining roadless areas because of their spiritual and cultural value to the larger society. "There is just one hope of repulsing the tyrannical ambition of civilization to conquer every niche on the whole earth. That hope is the organization of spirited people who will fight for the freedom of the wilderness."

Reading Marshall's understanding of wilderness nearly a century after he wrote about it, we find much that seems naive and, to be generous, obliviously racially insensitive. His notion of wilderness is a projection of his own needs. But those needs were not shallow or nonexistent. His grasp of American history was unsophisticated. He was a product of eastern, establishment wealth. But from the Adirondacks to Alaska he experienced something profoundly moving, and he desperately sought to make sure that experience would be available in the future to anyone who sought it. In his way, Marshall was sensitive to an American obligation to address historic inequities. He never surrendered his commitment to democratize the wilderness, striving to ensure that the joy he felt in the Adirondacks, Alaska, and countless other American wild lands would someday be accessible to all Americans, of any color.[29]

Marshall's sense of the wilderness was almost completely a function of his notions of adventure, history, and discovery and was thus poignantly performative. But it was the product of honest, deeply felt convictions. Combined with his amazing vitality and his willingness (almost eagerness) to enter into wars of words, memoranda, articles, books, and bureaucratic hearings, his dedication to wilderness preservation helped to secure the wilderness legacy we enjoy today. He was not an original or self-reflective thinker or environmental philosopher. The constructed wilderness that thrilled him as a boy remained largely the same to the end of his life. Marshall's wilderness experience was simultaneously deeply felt, i.e., transformative, and performative.

8

Forever Kept as Wild Forest Lands

In my conversations with Bill Verner and while reading Roderick Nash, Robert Marshall, and all the splendid travel writing about wilderness in the Adirondacks, I was always thinking about the Forest Preserve. In other words, I was trying to make the jump from myriad forms of personal appreciation to the policy decisions involved in establishing and protecting wilderness. The critical feature of this story is Article 7, Section 7, of the constitution written by delegates elected to a state convention in the summer of 1894, approved by New York voters in November of that year. It reads thus: "The lands of the state, now owned or hereafter acquired, constituting the forest preserve as now fixed by law, shall be forever kept as wild forest lands. They shall not be leased, sold or exchanged, or be taken by any corporation, public or private, nor shall the timber thereon be sold, removed or destroyed."

Despite several attempts to alter this language, it is still in the state constitution, in exactly those words and with the exact same punctuation, although it was renumbered, becoming Article 14, Section 1, at the convention of 1938. It has been expanded many times, to provide for specific, precisely identified exceptions for small, precisely identified places: a town cemetery, a town dump, a ski slope, a highway, a power line, and others.[1] But its governing authority over a steadily growing Forest Preserve has remained precisely the same since it became operational on January 1, 1895. Adoption of this constitutional provision is one of two key moments around which this book revolves—the other is the decision in *Protect*.

There were two main drivers of the critical legislation and constitution writing of the 1880s and '90s. First was a fear of what would happen to New York commerce if the forests of the Adirondacks and the Catskills were destroyed. Profoundly influenced by a best-selling book, *Man and Nature: or Physical Geography as Modified by Human Action* (1864) by George Perkins Marsh, influential New Yorkers came to believe that if irresponsible logging and the fires that frequently followed it continued to denude Adirondack slopes, a reliable flow of water to the Erie Canal and the Hudson River could no longer be guaranteed. Marsh had shown that mountain forests are essential to a reliable watershed and that a failure to protect them can lead to endless cycles of flood and drought. The Erie Canal had made New York the Empire State, and even in the age of railroads the loss of it and the Hudson River as transportation arteries would have been catastrophic for New York business interests.[2]

In addition to anxiety about watershed, the second key concern of legislators, constitutional delegates, and voters in the 1880s and '90s was protecting Adirondack forests, peaks, and waters as a recreational destination. Promoters of the Adirondack wilderness as the perfect locus for the restoration of the modern soul beleaguered by the frenetic pace of urbanism and industrialism had advanced a widely—but not universally—shared sense that the Adirondacks were essential to the state's welfare in ways quite different from the importance of the watershed to commerce. After studying the primary sources of these developments for my entire adult life, my view is that the watershed argument carried slightly more weight, but the idea of the Adirondack wilderness as spiritual retreat was close behind it.

Taking concrete steps to do something about protecting Adirondack forests did not come easy. After the Civil War, a slowly growing constituency began to ponder whether or not the State should intervene, but it took many years to move past the inertia of the status quo. In 1872, for example, the New York legislature appointed a citizens' committee to look into the need for State action, and it recommended the creation of an Adirondack Park. But the legislature let the matter drop.[3] Nothing substantive occurred until the 1880s and '90s with the creation of the Forest Preserve and Park and the adoption of a new constitution.

The roots of the Forest Preserve can be traced to February 6, 1883, when the New York Legislature adopted a law that permanently altered the State's policy with respect to Adirondack lands.[4] Prior to the passage of this law, nearly all state-owned lands in the Adirondacks were for sale. After the American Revolution and the establishment of New York as a state, the sale of publicly owned land in the state's undeveloped corners had been a source of income for the state government.[5] The legislature decided that whatever lands the state of New York still owned in 1883 in ten northern New York counties (Clinton, Essex, Franklin, Fulton, Hamilton, Herkimer, Lewis, Saratoga, St. Lawrence, and Warren) were no longer for sale. These lands, scattered across a huge swath of northern New York, constitute the origin of the Adirondack Forest Preserve. No explanation of the need for this law appears in the statute.[6]

That the legislature was keenly aware that further State intervention was necessary became clear the following year, when the legislature appropriated $5,000 and instructed the comptroller to appoint a commission to "investigate and report a system of forest preservation."[7] On July 3, 1884, Comptroller Alfred O. Chapin appointed Charles S. Sargent, a professor at Harvard and "a trained and eminent specialist" to work on such a report. Also appointed were D. Willis James, a New York City businessman; William A. Poucher, a lawyer from Oswego; and Edward M. Shepard from Brooklyn, also a lawyer. The appointment of such a commission has long been a feature of New York governance: the legislature or the governor identifies an issue or problem and selects a panel of experts to research the matter and submit a report with recommendations. (The key Adirondack example of this process is Governor Nelson Rockefeller's appointment of the Temporary Study Commission on the Future of the Adirondacks in 1968. This commission reported its recommendations, the main one being the establishment of the Adirondack Park Agency, in 1971.)[8] The report, known ever since to the few people who have read it as the Sargent Report, is a critically important and often overlooked document in the history of conservation and wilderness in the Adirondacks. It demands a thorough and close examination.

The commission appointed by Comptroller Chapin was described by him to be "competent in its administrative equipment, as well as in technical proficiency." It conducted its "work actively and zealously without pay." On January 23, 1885, the commission submitted its report to Chapin, who in turn submitted it to the legislature.[9] The Sargent Report is the first of a series of critical state documents that this book examines to establish the context for the decision in *Protect*. Among other things, the Sargent Report is the point of origin for the expression "wild forest lands."

The problem, how to efficiently and wisely use forests, noted Chapin, was not solely a New York concern. It affected the welfare of many states and the nation as a whole. "It is eminently fitting that in its solution the Empire State should lead the way." New York authorities reasonably believed that the state would be a leader when it came to forest conservation.

The commissioners examined forests in the Adirondacks and Catskills and heard testimony from "a large number of persons interested, directly and indirectly, in forest property." They sent experts to the Adirondacks to examine the forests firsthand. Their final report was accompanied by photographs showing "the condition to which excessive forest devastation has already reduced large areas within the water-sheds of the principal streams of the State." The Sargent Commission asserted that damage to the watershed, the capacity of rivers and streams flowing out of the Catskills and Adirondacks to provide reliable water to rivers and canals outside those two regions, was their primary concern.

At the time of their report, the commissioners found that the State owned 750,000 acres of "unimproved or forest land" in the Adirondacks (defined as "Northern or Adirondack counties") and 31,000 acres in the Catskills (in Delaware, Greene, Sullivan, and Ulster counties). In the Adirondacks, which, the commissioners announced, "has claimed the[ir] principal attention," the State possessed large holdings in Essex, Hamilton, and Herkimer counties, along with many "small and often widely-scattered tracts, rarely exceeding a few hundred acres in extent" throughout northern New York. The commissioners admitted that the precise extent and location of state lands was impossible to establish with certainty: survey lines, which were occasionally a century old, were "generally inaccurate or defective." Some titles were disputed by adjacent landowners. (Beginning in 1883, or the year of the legislation withholding all state land from purchase, determining boundaries between state and private land in the Adirondacks became the responsibility of Verplanck Colvin, and the name of his project changed from Adirondack or Topographical Survey to State Land Survey.)[10]

Because titles and boundaries were such a jumble, the commissioners declared that they could not examine only state land but rather would study "the forests without regard to ownership." They found, moreover, that while some state-owned lands in the Adirondacks were "still covered with the original forest growth, from others all merchantable timber has been removed, while there are considerable areas of these lands entirely stripped of forest covering."

That last sentence demands scrutiny. It is the first instance in the documents examined for this book of the expression "merchantable timber." The precise meaning of the word "timber" arose as a critical issue in *Protect*, and it's worth noting that the Sargent Report suggests that the word "timber," unmodified, means more than "merchantable timber." That is, to clarify that they were discussing removal of logs of market value, the commissioners apparently found the word "timber" inadequate and chose to modify it to clarify their meaning. This is a crucially important matter: the State and other advocates of removing trees to promote recreation, especially in the State's presentation of its position in *Protect*, have frequently and routinely argued that at the end of the nineteenth century, the word "timber" meant only trees large enough to provide logs of market value. This sentence in the Sargent Report challenges that assumption. To be sure, this one example proves little, but it is important to track such usages whenever they surface.

The Adirondack region, wrote the commissioners, was best suited for forests. Agriculture was marginal and generally unprofitable. "A wise economy" would understand that certain types of terrain are best left in a forested condition, but "the people of New York have not yet learned this lesson." Attempts to cultivate Adirondack soil "have resulted in disastrous failure, and abandoned homes and fields are scattered everywhere along the borders of the forest." Experience had shown that "the real and only value" of the Adirondack upland "consists in the forests which cover it." Echoing the admonition of George Perkins Marsh, whose 1864 *Man and Nature* forcefully argued that if upland forests did not remain intact, the result would be cycles of drought and flood in the valleys below, the Sargent Commission Report warned that removing Adirondack forests would "reduce this whole region to an unproductive and dangerous desert."

The Sargent Commission identified three critical values in Adirondack forests and listed them in descending order of importance:

First: Adirondack forests were essential to maintaining a reliable flow of water to the Erie Canal and rivers vital to New York commerce (among them the Hudson, the Mohawk, and the Black). Without water in these commercial arteries, the New York economy would collapse. Excessive logging on the Adirondack periphery, the report asserted, had already led to a thirty-percent reduction in the "summer flow" of rivers rising in the Adirondacks.

Second: The forests were essential to maintaining the recreational and aesthetic appeal of the Adirondacks. Tourism was rapidly becoming a vital regional industry. "Millions of dollars are invested in it; and thousands of persons draw from this business their principal support." Without intact forests, the Adirondacks would lose their appeal: they "supply the element of wildness [*sic*],—the charm which draws people to this region." With the forests destroyed, "this great lucrative business, capable of vast and permanent development, will be forever lost to the State." The Sargent Report thus emphasized the economic value of recreation and tourism. Forest destruction would mean that "the public will be deprived of the opportunity to enjoy the benefits which a visit to the Adirondack woods now offers."

Third: The Adirondacks supported another important industry, logging. "This business employs a considerable capital and a large number of men." Around the world, noted the commissioners, woods products were becoming scarce, and the laws of supply and demand inevitably meant that their "value in the future must increase." The proper course for New York was to design a "wise and comprehensive policy" of timber management, one that would generate profits for logging companies "without injury to the forests as reservoirs of moisture or as health resorts for the people."

The Sargent Commission thus emphasized commerce and multiple use. All three values attributed to forest cover—watershed protection, tourism, and logging—were important to the state insofar as they either protected commerce or generated jobs and profits. The notion of protecting wilderness as such, an unmanaged forest left intact for spiritual or ecological reasons, was not on the commissioners' collective mind. The Sargent Report was relentlessly utilitarian in its assessment of Adirondack conditions and its suggestions for State intervention.

After listing these three important contributions of healthy forests to the state, the Sargent Commission provided a lengthy assessment of the state's forests as they existed in the mid-1880s. The main point was that while much degradation had occurred, there was still time to ensure that the contributions of the forests to the state's welfare could be permanently protected. But the need for timely action was pressing: without immediate steps to conserve forests, "the State will feel the effect in empty rivers and increased droughts. A vast region will be deprived of its only permanent source of wealth, and its inhabitants must gradually sink into a condition of great misery and dependence. Its future lies in the Adirondack forests; if these are destroyed, its prosperity will disappear forever." The emphasis remained utilitarian.

What was causing the degradation of the still-extensive forests? The commission identified several factors:

1. Flooding of forests under reservoirs.

2. The manufacture of charcoal "especially in the region adjacent to the shores of Lake Champlain."

3. Logging. But the commissioners noted that since the Adirondack forest was largely hardwood, since the best means of getting

logs to market was by river driving, and since only softwoods float
well on river drives, the removal of mature hardwoods constituted
only a minor threat.

4. Fires. This was the chief threat. Loggers harvesting softwoods,
for either lumber or pulp, commonly left piles of slash on the ground
(twigs, branches, bark, anything that wasn't valuable), which became
tinder for forest fires. "Fires are slowly and surely destroying the
Adirondack forests. Unless they can be stopped or greatly reduced in
number and violence, nothing can prevent the entire extermination
of these forests." Fires originated in several ways: carelessness with
campfires and around logging camps, sparks from locomotives, and
clearing for agriculture.

The fire threat in the Adirondacks was characteristic of the age, part
of a larger apocalypse that burned millions of acres in the United
States and Canada in the years between around 1870 and World
War I, especially in the forests stretching from the Great Lakes to
New Brunswick. Fire historian Stephen J. Pyne has noted that the
logging practices of this period, which he labels "the great era of holo-
causts," created an "unrivaled tinderbox of abandoned slash."[11]

As of the time of the Sargent Report, burned areas in the Adirondacks remained mostly on the periphery, "a desert belt around the
remnants of the Adirondack forest." But every year saw this belt wid-
ening inexorably toward the center. In addition to destroying timber,
fires degraded the soil cover, interfered with forest succession, and
produced brambles and scrub susceptible to more, quick-moving fires,
which in turn further impoverished and eroded the soil. The critical issue before the State, argued the commissioners, was preventing
fire. And without substantive state intervention, the problem would
only get worse, as railroads penetrated the interior and provided both
incendiary sparks and a means of transporting hardwood logs to mill
and market.

What did the Sargent Commission propose?

While State ownership of Adirondack forests seemed a logical solution ("the future permanence of these forests cannot be absolutely

insured except through State ownership"), the commissioners considered this unlikely, at least for the present. It would be politically difficult, would threaten the presumed inviolability of private property rights, and would evoke strenuous opposition to any state interference in market forces. In other words, the public was not ready for exclusive state ownership of Adirondack forests. The commissioners further lamented that it would not be possible for the State to regulate logging on private land. After considering all the objections to state purchase—despite their certainty that "absolute control can be insured only by absolute purchase"—the commissioners ruled it out: "The enormous expenditure that would be required, and the danger of artificially enhancing the value of such lands for the sake of a sale to the State, have convinced the Commissioners that the State cannot wisely enter upon any scheme of general purchase."

Rather than embark on an extensive purchase program, the commissioners recommended that the State hold on to the Adirondack lands it already owned ("no inconsiderable part of the Adirondack forest") and acquire further lands as logging companies harvested softwoods and then opted not to pay real property taxes, thus relinquishing their lands to the state. A critical element in this plan was the recommendation that, on the lands owned by the state, "a wise system of forest management" be undertaken. This would demonstrate to the public that state ownership could be efficient and could contribute to the state's welfare. Once the state had shown itself to be a reliable and productive steward of its lands and that logging would continue to provide jobs, "purchase of other lands of the same character can be seriously considered."

As soon as boundaries were accurately established and policed, theft of timber from state land could be thwarted. And as the State gradually acquired more land from tax sales and as logging concerns began managing their lands more conservatively (how this would be promoted was not explained), protection of the state's rivers, the sole "aim and excuse for forest ownership by the State," would be secure. A further gain of this accomplishment would be "abundant opportunity for the enjoyment by the people of the pleasures and benefits of sylvan retreat."

In addition to figuring out exactly where its lands were and what their acreage was, the State should establish "a well-considered system" for their management. Decisions on what to do with public land should be removed from the purview of politics and be executed by "thoroughly trained, skillful and enthusiastic officials." It is important to note that among their responsibilities would be "the collection of some revenue from sales of timber which is cut for the better preservation of the remaining forests." This confirms that the Sargent Commission anticipated that timber would be harvested on state-owned forests in the Adirondacks but that the primary goal of such cutting was not so much the generation of revenue (though that was welcome) as the health of the forest itself.

Provision should be made, moreover, "for taxation of State lands within the forest preserve." This has two important features: first, it acknowledges at the moment of origin for the Forest Preserve that the State should pay taxes on its lands. Second, insofar as I know, it is the first instance in an official document of the phrase "forest preserve" as a descriptor for what state lands would be called.

Who could be relied on to make such decisions with skill and honesty? This person or persons must be conversant with both business and science, must be above politics, must above all bear in mind the interests of the state. They should not be elected officials as that would subject them to the whims of partisan politics. The answer, wrote the commissioners, was a Forest Commission.

The Sargent Commission recommended the appointment of an unpaid, "non-political" Forest Commission, appointed by the comptroller for terms of service long enough to ensure continuity and consistency of policy decisions. The Sargent Report explicitly assumed that only men (i.e., not women, who at that time could not even vote) would be chosen for this commission and that salaries "would bring an element of instability into the composition of such a commission, and sooner or later defeat the purpose for which it was created": instead, the men "would find abundant honor in preserving the public forests." The Sargent Commission thus introduced an element of elitism into the Forest Preserve at the moment of its establishment: the only people suitable for service on the proposed Forest Commission were wealthy men. The chief employee of the Forest Commission would be the forest warden, who would execute policies, be trained in business and science, and be a man of "tact, vigor, firmness and the breadth and aptitude of mind which are needed at the inauguration of any new policy."

The Sargent Commission devoted considerable attention to the matter of the State paying taxes on publicly owned forests. State ownership of forest lands was an issue of value to the State as a whole, and the State should therefore not burden local government by removing extensive lands from the tax rolls. By 1883, the year the State stopped all further sales of state lands, the State owned about 750,616 acres in the Adirondack region. The commission also recommended that all logging be forbidden on any private land on which tax payments were not up to date; this would prevent the practice by which forest owners acquired land, stripped it of value, and abandoned it before tax auctions could catch up with them. Insofar as I can tell, this is the first instance wherein the State revealed an intent to regulate, in any way, what occurred on private land in the Adirondacks.

The commission included in its report a series of bills, which it recommended be passed as soon as possible. The first of these, "An Act for the preservation and care of the Adirondack Forest," included these words: "The lands now or hereafter constituting the forest preserve shall be forever kept as wild forest lands. They shall not be sold or leased except as hereinafter provided, or taken or used by any public officer, or by any person or corporation, public or private, except as herein especially provided." Much of this language would eventually find its way to the constitution written in 1894. The Sargent Commission Report is thus the origin of the expression "forever kept as wild forest lands," which has been used ever since in connection with the New York State Forest Preserve and from which we derive the handy shorthand, "forever wild." The commission recommended that state lands could not be seized by eminent domain, that no lake or river in the Forest Preserve could be "altered, diverted or interfered with."

The putative duties of the Forest Commission were spelled out in considerable detail. It would manage state lands and encourage forest growth in the Adirondacks. A key responsibility of the Forest Commission would be supervising the harvest of timber sufficiently mature that it would be "advantageous for the general preservation of the forest." When it declared that it would "be forever kept as wild forest lands," this meant that the Forest Preserve would be wooded land, that it would not be a town or a farm. Nowhere in this report was there any expectation that the Forest Preserve would be untouched wilderness; this word does not appear in the Sargent Report. Further evidence that wilderness, as we now use the word, was not in the minds of the Sargent Commission was its explicitly stated understanding that the new Forest Commission could lease small parcels under its jurisdiction for "lodging-houses or hotels."

9

The Forest Preserve

On May 15, 1885, the New York legislature responded to the Sargent Report and effected final passage of Chapter 283 of the Laws of 1885 (hereafter referred to as Chap. 283, 1885, or simply the Forest Preserve law), translating the recommendations of the Sargent Commission into state law. This law created the Forest Preserve and the Forest Commission.[1]

The first order of business for the new law, which followed closely the Sargent Commission Report, was the creation of the Forest Commission. This was the one of the first New York State agencies with an explicitly environmental mission. Over the decades it has gone through many name changes and massive expansions of its authority and responsibilities. Along with the Fisheries Commission (1868), it was the distant point of origin of today's New York State Department of Environmental Conservation (hereafter DEC). A suggestion of the bureaucratic future of the Forest Commission can be found in the fact that its duties included, in addition to overseeing the actual Forest Preserve in the Adirondacks and Catskills, "the public interests of the state, with regard to forests and tree planting, and especially with regard to forest fires in every part of the state." The Forest Commission was to work with the regents of the State University to promote forestry in "the public schools, academies and colleges of the state." Another duty was the promotion of scientific forestry on private lands and planting trees on denuded public or private lands.

Where the Sargent Commission had recommended that the Forest Commission be appointed by the comptroller, the new law assigned this role to the governor, with "advice and consent of the senate."

Commissioners would be appointed by the governor and serve at his pleasure. To ensure continuity from one governor to another, their terms would be for six years. The commissioners would not be paid. The budget of the Forest Commission, which would be established by the legislature, would cover salaries of its employees: a "forest warden, forest inspectors, a clerk and all such agents, as they may deem necessary."

Section 7 created the Forest Preserve: "All the lands now owned or which may hereafter be acquired by the state of New York, within the counties of Clinton, excepting the towns of Altona and Dannemora, Essex, Franklin, Fulton, Hamilton, Herkimer, Lewis, Saratoga, St. Lawrence, Warren, Washington, Greene, Ulster, and Sullivan, shall constitute and be known as the forest preserve." There were eleven Adirondack and three Catskill counties in this originating legislation. Oneida (Adirondacks) was added in 1887, and Delaware (Catskills) in 1888.[2]

Section 8 further defined the Forest Preserve: "The lands now or hereafter constituting the forest preserve shall be forever kept as wild forest lands. They shall not be sold, nor shall they be leased or taken by any person or corporation, public or private." Sections 7 and 8 thus carefully followed the language of the Sargent Commission. They continued, moreover, the Sargent Commission's emphasis on acreage, not trees. As in the Sargent Report and as can be seen in other sections of the law, "wild forest" meant wooded land that was neither town nor farm. It did not mean wilderness in the modern sense.

The inalienability of the Forest Preserve—i.e., the fact that its lands would remain forever owned by the state and could not be sold or otherwise transferred to any other entity—has its probable point of origin in the way the state viewed its canals. Article 7, Section 7, of the then-current state constitution (New York's third constitution, drafted in the summer of 1846 and approved by voters in November of that year) stipulates: "The legislature shall not sell, lease, or otherwise dispose of any of the canals of the state, but they shall remain the property of the state, and under its management forever."[3] The similarity of the language protecting the canal system to that protecting state-owned forests deemed essential to the viability of the canals confirms the utilitarian, commercial impulse behind the origin of the Forest Preserve and the close association of the Forest Preserve with the state's canals.

The police duties of the Forest Commission included dealing with trespass and timber theft. In an ambiguous phrase critical to our understanding of twentieth-century litigation (especially of *Protect*), the statute specified among the kinds of trespass to be prosecuted "any act of cutting or causing to be cut . . . any tree or timber standing within the forest preserve . . . with intent to remove such tree or timber, or any portion thereof, from the said forest preserve." This suggests that the authors of Chap. 283, 1885, saw a distinction between "tree" and "timber." Or was the use of both these words an example of the sort of calculated redundancy one often encounters in legal writing aiming to minimize ambiguity or misinterpretation?

The new law foresaw income-generating logging to be a legitimate activity on the Forest Preserve. All revenue was to go into the state treasury. Finally, the law laid out extensive guidelines for addressing the statewide and pressing issue of fire: mainly recruiting and paying firefighting forces and the enforcement of regulations designed to minimize fires caused by locomotives.

The Forest Preserve law of 1885 law thus repeated the utilitarian thrust of the Sargent report. "Wild forest lands," words that would take on increasing significance as they were subsequently written in the constitution of 1894, did not mean, at this point, wilderness. Wilderness was not on the collective legislative mind. But this began to change in 1894.

10

The Adirondack Park

In *Wilderness and the American Mind*, Roderick Nash places special emphasis on New York's establishment of the Adirondack Park, which the state legislature created, at the urging of Governor Roswell Flower, in 1892. The legislation appeared to give equal value to protecting watershed and preserving "ground open for the free use of all the people for their health and pleasure."[1] The exact title of the legislation was "An ACT to establish the Adirondack park and to authorize the purchase and sale of lands with the counties including the forest preserve."[2]

Section 1 declares, "There shall be a state park established within the counties of Hamilton, Herkimer, St. Lawrence, Franklin, Essex, and Warren, which shall be known as the Adirondack park, and which shall, subject to the provisions of this act, be forever reserved, maintained and cared for as ground open for the free use of all the people for their health or pleasure, and as forest lands necessary to the preservation of the headwaters of the chief rivers of the state, and a future timber supply."

This language raises many questions. It suggests that the legislature at that point understood the Forest Preserve and the new Park to be one and the same. That is, the current situation where the Park includes both the Forest Preserve and privately owned land was not anticipated. It also appears to prioritize the recreational value of the new park over its watershed-protection function. This was a significant change from the 1885 Forest Preserve law. Finally, it explicitly declares that timber harvest would be permitted.

For the most part, the statute went on to stipulate the terms under which the State could purchase additional land and thus enlarge both the Forest Preserve and the Park. Consolidation and expansion of state-owned land, regardless of whether it was called Forest Preserve or Park, was the primary concern of the 1892 Park law. To this end, this law declared the appropriateness of the State's selling Forest Preserve land in cases where it did not fulfill the stated purpose of the new Park. This provision thus reversed the prohibition on selling state-owned land that began in 1883 and was repeated in the Sargent Report and the 1885 Forest Preserve law. Under the terms of the Park law, the income from any such sale would be earmarked for purchasing additions to the Forest Preserve.

Section 8 of this law confirmed the notion that the new state park was comprised only of state land, that, in other words, the legislators were not anticipating a mix of Forest Preserve and private land. Repeating a familiar formula, it asserted, "All lands now owned, or which may hereafter be acquired by the State within the towns mentioned in section two of this act (except such lands, in border towns, as may be sold in accordance with the provisions of section four) shall constitute the Adirondack park." As with the Forest Preserve, as defined in the statute of 1885, the new park was to be administered by the Forest Commission, which continued to have the authority to lease Forest Preserve tracts of up to five acres for five years for "camps or cottages." Section 10 repeats, "The Adirondack park shall for all purposes, be deemed a part of the forest preserve."

The explicit declaration of state commitment to enlarge the Forest Preserve through purchase and the implicit prioritization of recreation over watershed protection are important, but the 1892 Park law ignores the ongoing status of privately owned land inside the designated counties and the existence of dozens of towns and villages with thousands of residents. It thus casts a broad and murky ambiguity over any relationship between private land and the Forest Preserve. But the blue line, which indicated on official state maps the precise Park boundary described in Section 2, announced clearly the area that the legislature was interested in. It was significantly smaller than the Park of today, but it was another step in what has become the continuing, though halting, effort of the State to enlarge and protect its holdings in the Adirondacks.[3]

Following the passage of the 1892 Park law and aiming to reconcile the existence of a Park and a Forest Preserve, which were similar and to a large extent overlapped but were not identical, the legislature in 1893 enacted legislation to begin the sale of timber, to permit the sale of isolated Forest Preserve parcels (i.e., those not within the bounds of the new Park), and lease campsites.[4] This law declared, "The forest commissioners may sell any spruce and tamarack timber, which is not less than twelve inches in diameter at a height of three feet above the ground, standing in any part of the forest preserve, and poplar timber of such size as the forest commission may determine and the proceeds of such sales shall be turned over to the state treasurer, by whom they shall be placed to the credit of the special fund established for the purchase of lands within the Adirondack park."

This language confirms two critical issues. First, the legislature continued to understand "wild forest lands" to mean something other than wilderness in the modern sense of the word: limited logging was permitted on the Forest Preserve. Second, as in the Sargent Commission Report, the word "timber" needed modification to establish clarity. In other words, "timber" did *not* exclusively mean only trees containing marketable logs. In order to prevent the cutting of smaller trees, the legislature saw the need to indicate a specific diameter at slightly below breast height.

The law went on to reconfirm (in section 106) that the State would pay taxes on Forest Preserve lands. Most of the law involved provisions to prevent and respond to forest fires. It also provided for the State to offer five-year leases for "camps or cottages."[5] While adding little to our understanding of the current understanding of the definition or purpose of the Forest Preserve, the Park law and the legislation of 1893 illustrate continuing concern about the ostensibly incomplete status of the protections in place and the State's interest in all the lands inside the blue line.

11

The Constitution, 1894

A convention to rewrite New York's Constitution gathered in Albany in the early summer of 1894. Since the previous convention, held in 1846, New York had become the center of the nation's financial system, had evolved into a manufacturing and transportation colossus, and had added millions of immigrants to its population. It had many issues to address—the rules governing apportionment of legislative districts, the always-complex machinery of New York's court system, education, and the power relationship between the state government in Albany and local municipalities, among many others.[1]

Not high on most delegates' list was the Forest Preserve, but the Forest Preserve did have one key advocate, the New York Board of Trade and Transportation (NYBTT), apprehensive about the reliability of water in the Erie Canal and the Hudson River. The NYBTT was established in 1873 "to promote the Trade, Commerce and Manufactures of the United States, and especially of the State of New York."[2] In 1893 the secretary of the NYBTT, Frank Gardener, increasingly alarmed by the inadequacy of the 1885 Forest Preserve law and by efforts of the legislature to weaken it even further, declared, "I am convinced that the forests will never be safe until they are put into the State Constitution."[3] Immediately before the convention opened its deliberations, the NYBTT issued a pamphlet, "Proposed Amendment to the Constitution of the State of New York to Preserve Its Forests, with Reasons Why," describing the weaknesses of both the 1885 Forest Preserve law and the 1892 Adirondack Park law and arguing that no measure short of a provision for forest preservation in the state constitution would protect the watershed vital to the Erie Canal and the Hudson River.[4] In this pamphlet the NYBTT proposed a four-section provision to be added to a new constitution.

Throughout the discussion below, including everything that occurred before and during the convention and beginning with this pamphlet, we must pay special attention to the word "timber." That word added a critical expansion of the 1885 Forest Preserve law, which was concerned with land and the need to ensure that whatever land the State owned or might later acquire could never be alienated from state ownership and that it be covered with trees, some of which could be harvested should the State so choose. With addition of a provision that none of the timber on state land could ever be harvested, the Forest Preserve became more than land and its cover; from this point forward it was the land and *all* the trees thereon with which the State was concerned. And from this point forward the idea of the Forest Preserve as protected wilderness becomes a legitimate subject for our consideration, even though many (but not all) delegates to the 1894 convention probably did not have our understanding of that word in mind.

Deployed in the constitutional provision passed in 1894 to protect the Forest Preserve, "timber" has been the source of intense debate ever since, most recently in *Protect the Adirondacks! Inc. v. New York State Department of Environmental Conservation and Adirondack Park Agency.* The critical question is this: did the delegates in 1894 understand the word "timber" to mean all trees, little trees and big trees? Or did they understand the word to mean only trees large enough to be the source of marketable logs, for either lumber or pulp? Much depends on this distinction, and I will address it in detail. Doubtless, the meaning of "timber" varied among the many delegates: some understood it narrowly, some expansively.

Section 1 of the NYBTT's proposed amendment identified the Forest Preserve counties and echoed (but did not repeat verbatim) the language of the 1885 law: the lands of the "State of New York . . . shall be forever kept as wild forest lands, and shall constitute and be known as the forest preserve."

Section 2 stipulated that "they [i.e., lands] shall not be sold, neither shall they be exchanged for other lands, nor taken by any person or corporation, public or private; nor shall their woods be cut or sold, nor their downed or burnt timber removed; nor shall they be leased, except as provided in this article." Nothing else appears in Section 2. This is where the NYTBB manifested its certainty that the Forest Preserve law of 1885 was insufficient. Not only did the State need to own forever the land of the Forest Preserve, it needed to keep those lands forever covered with trees, none of which could be cut for any reason. With a prohibition on commercial logging (on state land), the NYBTT was demanding a radical departure from the 1885 Forest Preserve law. Throughout the remainder of the pamphlet, the words "forests" and "woods" are used interchangeably. Logging and fires posed a horrible threat to these forests and thus to the commercial health of the state.[5]

The remainder of the pamphlet is a lengthy explanation for why the NYBTT believed constitutional protection was called for, especially for why the forest cover was in danger and not adequately protected by the Forest Preserve law. It takes the form of a memorandum from the NYBTT to the convention, dated June 13, 1894. For the most part, it makes its case with an argument derived from Marsh's *Man and Nature*, which is explicitly quoted, about the importance of mountain forests to protecting a reliable and stable watershed. It lamented "the growing irregularity and uncertain flow of many of the streams that are fed by the natural reservoirs, the ponds and lakes, of these woods. More serious than this, some have lost half of their former volume, and others are completely dry."[6]

Among the immediately pressing reasons for demanding that protection of the Forest Preserve be added to the constitution was an 1893 law permitting the Forest Commission to sell "the standing spruce above twelve inches diameter three feet from the ground."[7] As the NYBTT observed, "The woodlands of the Adirondacks have been so thoroughly denuded of their soft woods in some sections, that the example of the State in seeking to dispose of such timber is not an assuring one to those who would keep her forests intact for the numerous and necessary benefits they confer."[8]

The NYBTT opposed selling off detached parcels outside the Park. In arguing for keeping these in state ownership, it notes, "Dead and decaying trees not only provide nourishment and soil for the young and growing timber and mosses but they cause the retention of much moisture."[9] This use of "timber" appears to include trees not yet of marketable size. Throughout this pamphlet, the argument for enhanced protection of Adirondack forests is relentlessly utilitarian. The concern for the recreational and spiritual assets of wilderness, which quickly became part of the floor debate, was added at the convention. The NYBTT initiated the utilitarian argument, but the convention itself advanced far broader concerns.

The Forest Preserve debate began on September 7, 1894. David McClure, a Manhattan attorney, and a member of both the Committee on Conservation and the NYBTT, introduced that committee's recommendation to the convention, a new constitutional article to read thus: "All lands of the State now owned, or hereafter acquired, constituting the forest preserves [sic] shall be forever kept as wild forest lands; they shall not, nor shall the timber thereon, be sold."[10] The NYBTT's proposed amendment had passed muster with the Committee on Conservation and had been carefully reworded for presentation to the convention as a whole.

David McClure is an elusive figure. He was the point man for the NYBTT at the convention, but we do not know whether he ever set foot in the Adirondacks. Donaldson declares, without documentation, that McClure was a Democrat at a convention where Republicans were in the majority and that his "strong sympathies with the forest movement were well known."[11] A brief obituary in the *New York Times* reveals only that he was born in Dobbs Ferry in 1848, was admitted to the Bar in 1869, spent his career as counsel for various banks and insurance companies, and was a trustee of the Roman Catholic Orphan Asylum. He died on April 30, 1912, at his home on West 49th Street. Governor David B. Hill (whose term as governor was 1885–91) offered him an appointment to the New York State Court of Appeals, but

McClure opted to remain in private practice.[12] The offer of such an important judicial position suggests that McClure was a prominent and well-respected attorney in the business community. That the NYBTT asked him to represent it at the 1894 convention seems a logical and shrewd choice. When I think of McClure working and socializing in Manhattan, I picture an expensively dressed and meticulously manicured character from a novel by Edith Wharton. That he traveled in New York City's elite social circles is confirmed by his and his wife's inclusion in the New York City Social Registers of the 1890s.[13]

At the convention, McClure immediately asked unanimous consent to amend the Forest-Preserve provision as referred from committee to read thus: "The lands of the State now owned, or hereafter acquired, constituting the forest preserves [*sic*], as now fixed by law, shall be forever kept as wild forest lands; they shall not be sold, or exchanged, or be taken by any corporation, public or private, nor shall the timber thereon be sold."[14] What was "woods" in the NYBTT's proposed amendment had become "timber." After this, the operative word in the Forest Preserve provision, was "timber." But in debates and speeches, delegates frequently substituted "trees" or "woods" for "timber," and most of them used these words interchangeably.

Beginning the next morning, McClure spoke at length on why the Forest Preserve was in need of enhanced protection. He described the state's holdings and quoted a long speech "made a few years ago by the Hon. George W. Smith of Herkimer," focusing almost exclusively on the Adirondacks and emphasizing the watershed argument.[15] "What is the value of these woods and why should we try to preserve them intact?" asks McClure. "First of all, because they are the woods—and in passing let me say, this State, strange as it may appear, has a very small proportion of its area in forest lands." He then develops two main threads to his argument: the first, interestingly, is the importance of the Adirondacks as a "great resort for the people of this state," thus expanding significantly the argument for protection expressed in the NYBTT's pamphlet. He frequently refers to "the woods," suggesting that the entire forest ecosystem, not just the marketable timber, demands protection. His second thread is the familiar watershed argument and the contribution of mountain forests to commerce.[16]

To emphasize the importance to the general welfare of the state of the entire forest, not merely the larger trees, McClure declares,

> We should not sell a tree or a branch of one. Some people may think in the wisdom of their scientific investigations that you can make the forests better by thinning them out and selling to lumbermen some of the trees, regardless of the devastation, the burnings and the stealings that follow in the lumberman's track. But I say to you, gentlemen, no man has yet found it possible to improve upon the ways of nature.[17]

"A branch of one"? This is surely a call to protect far more than trees offering logs of market size. He is talking about an entire ecosystem, with all its constituent parts, large and small, though of course the word "ecosystem," a twentieth-century neologism, does not appear in his remarks.[18]

Since the 1894 Constitutional Convention and its deliberations are so critically important to the history of the Forest Preserve and particularly to its history in the courts, leading up to and including the decision in *Protect*, we need to look carefully at what the delegates other than McClure had to say about his proposal. What follows is a tour of the arguments offered before the final vote on what became Article 7, Section 7, of the New York Constitution:

Delegate Goodelle moved to amend the proposal to read thus: "[N]or shall the timber thereon be sold, removed or destroyed," adding "removed" and "destroyed" to what would be prohibited with respect to the timber. He explained that this was to prevent the State from flooding parts of the Forest Preserve for reservoirs. Goodelle then confirmed the importance of the Adirondacks as a recreational destination.[19]

Delegate Floyd proposed that the State be permitted to exchange Forest Preserve parcels outside the blue line for private land inside the blue line and to sell firewood to local residents.[20] The suggestion about firewood was eventually voted down. As historian Karl Jacoby has effectively shown, the refusal of the convention, where very few of the delegates actually lived in the Adirondacks, to permit local residents to heat their homes, businesses, and churches with firewood from state land, was considered by those residents to be early evidence that Adirondack matters would be decided largely by outsiders.[21]

Delegate Charles Mereness declared that the Forest Preserve's "incalculable value to the commercial and material interests of the state as a natural reservoir of pure water" was of "far greater importance" than its use as a "health resort," but if we judge from how he pitched his remarks, it's clear that what he valued was wilderness. He said that he had "traversed this great forest in true aboriginal style, with my boatman and skiff, and a pack on my back." He had camped and explored where "none of the ravages of man were discernible." He feared that "the whole region will be made desolate and barren, unless the hand of the despoiler is stayed. We have already waited too long, but I implore you, do not longer hesitate to take measures to stop this outrage." Mereness acknowledged that the State might at some point in the future need to harvest its own forest resource but that this eventuality was at least fifty years off.[22]

Delegate McIntyre opposed any land exchanges.[23]

Delegate McArthur spoke of personal experience with declining water levels in the Hudson and said he had recently walked across the river at Fort Edward without getting his feet wet.[24]

Delegate E. A. Brown spoke in favor of strong protections and condemned the Forestry Commission's wish to sell standing timber to cover its expenses:

> I say, sir, that it is necessary that something be done; I say, sir, it is necessary to close the door unless you want this great water supply, this great sanitarium, this great health resort of our State that is known from ocean to ocean, and from land to land, destroyed, that you must shut the door, and you must close it tight, and close it right away; and not only that, you must keep it closed for twenty years.

Brown appeared to be more concerned with the State's incompetence or even corruption than with preserving the wilderness for recreational or spiritual reasons. He also argued that the state should own all of the Adirondacks, so that everyone, not just the rich, could enjoy the Park.[25]

McClure objected to all proposed changes, except the one adding "or destroyed." After some give and take, McClure declared that his committee's final proposal to the convention should read thus: "The lands of the State now owned or hereafter acquired, constituting the forest preserves [*sic*] as now fixed by law, shall be forever kept as wild forest lands. They shall not be leased, sold, or exchanged, or taken by any corporation, public or private, nor shall the timber thereon be sold, removed or destroyed."[26]

Any deviations from this, as proposed by the speakers other than McClure (i.e., those that would have changed McClure's text), were voted on and rejected, and McClure's language was finally approved on September 13, 1894. It was the only new constitutional provision to be unanimously adopted.[27] By the time this provision was voted on by the people, the plural "forest preserves" had become "forest preserve." No indication in the published proceedings of the convention's deliberations provides any clue as to how and when this occurred. I suspect that the plural was originally deployed because the Forest Preserve existed both in the Catskills and the Adirondacks and that somewhere along the line McClure decided it wasn't necessary.

12

Timber

What did the delegates of 1894 mean to protect when they voted to add the words "nor shall the timber thereon be sold, removed or destroyed" to the final version of Article 7, Section 7? What was meant by the word "timber"? One common understanding today, which the state of New York promoted in *Protect*, is that "timber" meant only trees large enough to provide logs of merchantable value. The State insisted that "timber" did not include immature, small-diameter trees, or any tree that could not furnish logs big enough to generate a profit for anyone who harvested them.

In its brief before the New York State Court of Appeals in *Protect*, the State declared,

> The historical context surrounding enactment of the forever wild provision, DEC's long-accepted use of the three inches dbh [diameter at breast height] standard, and other New York statutes and regulations addressing tree cutting demonstrate that the term "timber" in the forever wild provision is properly interpreted to refer to merchantable trees—not seedlings, saplings, or trees smaller than three inches dbh.[1]

If we examine closely how the word was defined and used at the end of the nineteenth century, however, we can clearly see that the State's assertion is selective and incomplete. Like all words, "timber" is hard to pin down; the more we demand absolute precision, the more elusive it becomes. While it was often used to mean what the State says it meant, we can find significant evidence that it also, to some people, meant more than that. I am not arguing that "timber" meant, to all who uttered the word, at all times, every tree, large and small. Clearly, the State's definition reflected the usage of some delegates. But I *am* arguing that the State was wrong to argue that "timber" can *only* be "interpreted to refer to merchantable trees—not young growth or trees smaller than three inches dbh."

Webster's International Dictionary of the English Language was the definitive standard for American usage at the close of the nineteenth century.[2] Its first definition of "timber" reads thus: "That sort of wood which is proper for buildings or for tools, utensils, furniture, carriages, fences, ships, and the like;—usually said of felled trees, but sometimes of those standing." Right from the start, we encounter ambiguity: it includes wood used to make objects as large as a building and as small as a tool or utensil. "Timber" is thus not *only* wood large enough for a two-by-four or for planking a ship's hull. It is both a construction member *and* the shaft of a hammer, both a two-by-four *and* a wooden spoon.

The second definition reads "The body, stem, or trunk of a tree." Note: the size of the tree is not indicated. If we look for help in understanding the breadth of this definition, moreover, and turn to check how "stem" is defined in the same dictionary, we find "The principal body of a tree, shrub, or plant, of any kind; the main stock; the part which supports the branches or the head or top." In *An Historical Dictionary of Forestry and Woodland Terms*, we find the word "stem" explained thus: "In 1832 this was defined as: 'The body of a tree in all its stages of growth, from a seedling to that of a full grown tree.'"[3]

A further definition of "timber" in *Webster's* is "Woods or forest; a wooded land." This sense of "timber" to mean the woods, the forest as a whole, the complex nexus of trees and undergrowth, is a common usage, as shown, for example, in Headley's *Adirondack*. Describing the exploits of the famous guide John Cheney, Headley, recalls "his long stretches through the wood [*sic*], where the timber is so thick you cannot see an animal more than fifteen rods." Here Headley clearly intends "timber" to mean the characteristically tangled, mixed Adirondack forest, not a stately array of individual, tall, mature trees whose canopy stunts the growth of the understory.[4]

The sense of "timber" that emerges from contemporary usage thus includes but moves far beyond and embraces far more than the State's narrow and exclusive understanding. It is obviously impossible to know with certainty what any single delegate understood "timber" to mean in the summer of 1894, but it is clear that the State's insistence on trees of market value and nothing else is conveniently selective, ignoring relevant alternatives. It reflects a naive understanding of how language, which is unavoidably circular, self-referential, and nuanced, actually works.

A further illustration of how "timber" could mean far more than what the State insisted it exclusively meant in *Protect* can be found in a report from Adirondack Surveyor Verplanck Colvin. In 1895, the same year the constitutional protection of state-owned "timber" went into effect, Colvin was working around Ampersand Lake and learned of the plans of the Santa Clara Lumber Company to log high on the slopes of Mount Seward:

> The agent informed me that not only was the timber in the valley to be cut and removed, but chutes were to be constructed, far up towards the summits of the high peaks, so that not only logs fit for lumber could be sent down to the skidways, but even the small softwood spruce timber would be thoroughly cut for pulpwood.[5]

For two reasons, this is an important passage. First, it demonstrates with clarity how the word "timber" was far more inclusive than the State insisted in *Protect*. Colvin is referring to the krummholz, the dwarfed, gnarled growth near the tree line on a High Peak, trees valueless for construction but useful as pulpwood. This raises the second reason: by 1890, a major use of Adirondack logs was for making paper, not for lumber, while on the eastern edge of the Park, hardwoods were harvested to make charcoal. Delegates to the constitutional convention, particularly those from the northern part of the state, would have been familiar with the need for pulpwood for the manufacture of paper. When they set out to establish protection for the forests of the Adirondacks, for whatever reason, they knew that cutting trees for lumber, pulpwood, and charcoal had to be prohibited on state land.

When it came to pulpwood, the size of the tree meant very little indeed. Yet they found "timber" to be fully adequate to their purpose.

The constitution written in the summer of 1894, including the protection of the Forest Preserve as "wild forest lands" and the prohibition against any removal or destruction of "timber," was approved by voters in November and became effective on January 1, 1895. The Forest Preserve provision was a radical departure from the sorts of issues ordinarily addressed in a state constitution. How the state's forests and other natural resources should be managed had been an issue for the legislature and the new conservation bureaucracy. It had been, observed constitutional scholar and historian Charles Lincoln, a matter "primarily of legislative cognizance, and which ordinarily might be left to the discretion of the legislature." The legislature quickly tried to recapture its control of the Forest Preserve and effected first passage of a redrafted Article 7 in the legislative session of 1895. Second passage followed the next year, and the people were asked to decide in November of 1896 whether they wanted to abandon the unanimous intent of the convention of 1894. The new language would have returned to the legislature authority to sell or lease lands in the Forest Preserve. The people resoundingly said No, by nearly a two-to-one margin (710,505 votes against and 389,019 for).[6] This was the first of several major efforts to dilute or even eliminate the tight constitutional protection of the Forest Preserve.

Just why the voters preferred—twice, once in 1894 and again in 1896—to keep protection of the Forest Preserve under their own control, and not to trust the legislature or the state bureaucracy, is hard to tell. It is also difficult to know whether the people believed the constitutional protection was temporary, a firm guard needed only until the logging industry could be trusted to harvest trees conservatively, without the waste and fires that characterized Adirondack logging between the Civil War and the early twentieth century. We do know that some delegates at the 1894 convention expressed their assumption that the restrictions of forever wild would eventually be lifted. And it is clear that at least some in the state, including, for example, Governor Frank Black, who declared to the legislature in 1897 that the paper industry was a vital part of the state's economy and that eventually the prohibition would be "changed," shared this assumption. Once the logging industry had "learned prudence instead of waste, and have equipped themselves with knowledge and experience adequate to the care of this great domain," the constitution could be amended.[7] It remained for the constitutional convention of 1915 to settle this matter.

13

Forever Wild

I left the Adirondack Museum and Long Lake in the early summer of 1973 and started teaching English and Latin at Episcopal High School the following September. I soon missed the Museum and the Adirondacks. After two years at EHS, I applied for and was admitted to the graduate program in American Studies at George Washington University, a short hop across the Potomac from Alexandria, Virginia, where I lived. That department had an intriguing working relationship with the Museum of American History at the Smithsonian, and I matriculated with the assumption that I would concentrate in American material culture and eventually navigate my way back into the museum world.

American Studies as a distinct degree, something different from American history, evolved at a handful of American universities, beginning with Harvard, in the 1930s. It reflected the wish of certain scholars to contextualize American literature, to study our classics as representative and thus illustrative of the culture that produced them, rather than as puzzles with symbols, metaphors, and other formalistic elements to be interpreted by a properly trained critic, i.e., someone with a PhD in English. Most American English departments of the day were dominated by what became known as the New Criticism, a methodology mostly focused on style and technique; it largely divorced literature from the biographies of authors and the political or cultural conditions of the era in which it was written. This certainly was the primary emphasis at Princeton when I was an English major there in the 1960s. (This is a sweeping generalization: there were plenty of exceptions.)[1]

American Studies originated as a revolt against orthodoxy. Scholars like Perry Miller at Harvard set out to study literature in a different way, with an approach more attuned to intellectual history. They also wanted to expand what was considered literature to include far more than just the familiar canon of fiction, poetry, and drama, nearly all written by White men: Puritan sermons, dime novels, popular verse, exploration narratives, previously ignored works by writers of color—pretty much anything that found its way into print—could be useful in understanding the cultural climate that produced it. It was but a short step to start studying literature, in the broadest sense, alongside paintings, architecture, and all that we now think of as material culture. American Studies was, in what remained its identifying claim to legitimacy for decades, "interdisciplinary."

The golden age of American Studies' marriage of literary study with cultural or intellectual history resulted in what became known as the myth-symbol school, of which Nash's *Wilderness and the American Mind* was the example that first caught my attention. The term "myth-symbol" derives from one of the most brilliant and innovative of this genre's offerings, Henry Nash Smith's *Virgin Land: The American West as Symbol and Myth.*[2] Published in 1950 by Harvard University Press and an obvious influence on Roderick Nash (they were not related), this book utterly seduced me when I first encountered it while pondering my options for graduate programs. Like *Nash's Wilderness and the American Mind, Virgin Land* is place-based. It asks what did people in nineteenth-century America think of the West? What was the popular understanding of the West's meaning as expressed in everything from dime novels to exploration narratives? Finally, how did these understandings, which might be wildly divorced from physical reality, come to influence policy? The issue of policy was key: can we understand how popular understandings about a place lead to policy, to decisions about what a place is good for, to how legislation concerning it is framed and executed? All of these elements combined to make *Virgin Land* a convention-breaking book, and they were all adopted by Roderick Nash in *Wilderness and the American Mind.*

When I first applied to GWU, I knew little about the myth-symbol school; what appealed to me was the fact that the American Studies Department offered a concentration in American material culture and that this was overseen by Wilcomb E. Washburn. His PhD, from Harvard, 1955, was in that institution's fledgling American Studies program, with the formal name American Civilization, the same departmental name adopted by GWU, one of the first schools outside the Ivy League to develop both undergraduate and graduate programs in American Studies. Washburn's official job was at the Smithsonian, but he also had a faculty appointment at GWU and routinely taught and advised graduate students working on material culture and angling for museum jobs. The relentlessly patrician Washburn possessed an apparently endless supply of crisp, oxford-cloth blue dress shirts, which he always wore with a preppy tie. Washburn and the Smithsonian, I thought, would be my ticket back into the museum world.

GWU required each PhD student to prepare in four fields of study, each of which culminated in an all-day written examination; all candidates in American Studies were examined in American cultural history, while one of the four fields had to be in a non-American subject. We each thus picked three fields to complement American cultural history. My first was material culture; my other two, picked because I had been an English major as an undergraduate and not because they made any sense for someone aiming for a museum job, were American fiction and nineteenth-century British literature. Even though I was in fact duly examined by Washburn in American material culture (I passed), by the time my preliminary exams were behind me and I was looking for a dissertation topic, I had drifted back toward the roots of American Studies, the intersection of literature and history, largely pushed by my growing fascination with the myth-symbol school and its confident and, to me anyway, mesmerizing deployment of the sweeping generalization. I had wandered away from contemplating a career in the museum world and hoped to become a college professor.

My preliminary exams were finished in 1977, and I needed a dissertation topic. After some early fumbling—which baffled my grad-school friends, who had been listening to me limning the wonders of the Adirondack wilderness for three years and could clearly see that my dissertation should be on some Adirondack theme—and with a critical and career-making prod by one of my professors, Joanna Schneider Zangrando, I settled on the idea of using Nash as a model and applying his reading of a broad sweep of American culture to the specific case of wilderness in the Adirondacks.[3] How, my dissertation asked, did the case of the Adirondacks illustrate or differ from the attitudes toward wilderness described by Nash?

Once settled on a subject, I had a significant advantage not shared by the others in my cohort. Because of my research at the Adirondack Museum, I was already familiar with much of the primary literature. The report I had written about wildlife depended heavily on the literature of exploration, the sporting and travel narratives of Headley and others, the Colvin reports, and the voluminous, fascinating, and profusely illustrated documents of the New York conservation bureaucracy. All these would be major sources for figuring out what explorers, hunters, anglers, and the state government itself thought about wilderness. Nineteenth-century sporting narratives were probably the most important vehicle by which an image of the Adirondack wilderness was constructed. It was largely because nearly all of my research at the Museum ended with the turn of the twentieth century that I concluded my dissertation with the constitutional convention of 1894.

I secured a carrel in the GWU library and worked there nine to five, five days a week. The Library of Congress (LC), with a collection housing almost every book, article, or document I needed, was a bus ride away. The LC became even closer when the Metro opened; there's a line running straight from the Foggy Bottom Station at GWU to the Capitol South Station across the street from the LC.

The routine for researchers at the Library of Congress involved submitting a slip of paper at a central desk; it listed author, title, and call number of whatever item I had identified as potentially useful via the card catalog, which in those days meant actual stiff-paper cards in the classic, long wooden drawers. When I began my research at the LC, I usually sat and waited for my books in the main Reading

Room of what is now called the Thomas Jefferson building. It was the first building for the Library after Congress determined that it was outgrowing the space allotted for it in the Capitol itself; it opened to the public in 1897. The Main Reading Room is a vast, magnificent domed space with giant columns and allegorical paintings and sculpture.[4] Sitting at one of the ancient wooden desks circling the center, I was inhaling the spirit of generations of scholars. But the ambient charm of the space quickly wore off as I waited endlessly for books that frequently failed to materialize, and my request slip was returned with the letters "N O S" scribbled on it. This meant "not on shelf"; in other words, the person designated to retrieve the book I requested couldn't find it. This happened repeatedly.

One day I walked across the street to the office of my congressional representative, Herbert Harris, and politely requested that I be given a stack pass. This turned out to be remarkably easy. I described to an administrative assistant what I was doing, told her about my difficulties, and asked whether Congressman Harris's office could help me. (I also said that I had voted for Harris, which was true, but she said that was irrelevant: he was there to help all his constituents.) She told me that an official clearance for a stack pass would be mailed to me in the next few days, and it was. I duly took this to an office at the LC where a secretary issued me a pass. More important, indeed crucially important, she gave me a map (no compass needed). Without it I would never have found anything. After this, I wandered at will along miles upon miles of metal shelves, mostly underground, on which sat millions upon millions of books belonging to all Americans. Always adding to the list of titles I needed to consult, I dipped routinely into the *Adirondack Bibliography*, a two-volume and utterly indispensable contribution to Adirondack research.[5] Throughout my time in the LC and the other collections I visited, I always kept both volumes of the *Bib* in my book bag.

At the LC, deep in the stacks, I would sit on the floor when I located a book I was looking for, write out notes or copy passages onto five-by-eight index cards, and then look at the books on either side of the one that had brought me to that corner of the LC's endless catacombs.

This often produced other items that I didn't know about and that proved to be useful. I also figured out why so many of my original request slips came back marked "N O S." The runners, the young and poorly paid employees whose job it was to fetch the books for all the eagerly waiting scholars, spent a good bit of their time socializing in the stacks while passing around joints. Given the monotony of their job, I think I would have done the same. Throughout the time that I wandered at will through parts of the LC ostensibly closed to the general public, no one ever challenged me or asked me to show my pass. In this age of security anxiety, I doubt that would happen now.

Researching and then writing my dissertation was one of the highlights of my academic life. And so was working with Barney Mergen, a young, recently tenured associate professor at GWU, for whom I had earlier written a long paper on nineteenth-century camping in the Adirondacks and whom I asked to serve as my director. Barney was a westerner, having grown up in Reno, Nevada, and was moving into the emerging field of environmental history at just the time I started at GWU. He liked the paper I had written on camping and suggested I deliver it at a scholarly conference. I did so, at an annual meeting of the North American Society for Sport History, conveniently scheduled nearby in College Park, Maryland. After my session—my first at such a meeting—the editor of the *Journal of Sport History* invited me to submit an article. After much revision and good suggestions from Barney, I did, and this became my first article published in a refereed journal.[6] It developed an important theme of the dissertation, a point where I departed from complete fidelity to Nash: in my reading of the responses to how the sportsmen I called Romantic Travelers—Headley and his contemporaries—responded to the Adirondack wilderness, I noted an ambivalence, a vestigial skepticism about wilderness behind the enthusiasm. Nash had found a nearly unalloyed appreciation, a key step on the way to the preservation advocacy of John Muir and the triumphant climax of the 1964 Wilderness Act.[7]

I reread everything I had digested at the Adirondack Museum, filling in many gaps around the edges. Roderick Nash's model dominated pretty much every step along the way, although I finished up my study with the New York constitutional convention of 1894. (I wanted to get my degree as fast as possible and enter the academic job market, knowing that if I were eventually to try to turn the dissertation into a book, I would need to expand it well into the twentieth century.) *Wilderness and the American Mind*, in its first edition, 1967, concluded with the passage of the 1964 Federal Wilderness Act. For Nash this act was the final, triumphant moment, signifying the complete, 180-degree turn away from Puritan hostility.

In contrast, I found some uncertainty at the 1894 Constitutional Convention. Indeed, I now think I exaggerated this, trying too hard to disguise a slavish application of the Nash model. I read the proceedings of the 1894 Constitutional Convention and perhaps overemphasized the watershed argument, the clearly expressed concern on the part of the delegates that irresponsible logging and the frequent and devastating fires that so frequently followed would destroy the reliability of the watershed. This was there, of course, but so was pervasive appreciation of the spiritual asset that the Adirondacks and the region's remaining wilderness offered to New Yorkers struggling with the fever of frenetic industrialization and the rapid growth of cities.[8]

Otherwise, my dissertation follows the Nash model faithfully. Where Nash was looking at the entire United States, I enjoyed the obvious benefit of a narrower scope. But what Nash did in approaching his subject was exactly the strategy I adopted. Deploying the myth-symbol model, I read exploration and travel narratives, government documents, surveyors' reports, and everything I could find where someone encountered the Adirondack wilderness in the nineteenth century and expressed in one way or another what he (or, rarely, she) thought about it. This is a methodology that requires finding the useful passage or sentiment, quoting it, and then showing how it advanced my thesis. It's what I was taught to do as an English major when I wrote analyses of various works of literature. It's a simple, well-worn, and more or less unavoidable way of developing a topic like mine (or Nash's). An inevitable danger of this method is that the historian can easily fall into the trap of cherry-picking the evidence, emphasizing the passages that say what I want them to say and ignoring or simply not registering when the source displays an attitude that contradicts my needs.

As Nash had done, I showed how these developing attitudes eventually led to policy, in other words, how a polity, the state of New York, translated widespread cultural values into statutory or other projections of state authority. In the Adirondacks, this meant the three critical late nineteenth-century moments: the establishment of the Forest Preserve in 1885, the designation of the Adirondack Park in 1892, and the adoption at the state constitutional convention of 1894 of what has been known ever since as the forever-wild clause. This is where the dissertation ended. I graduated in 1979.

Combined with the course work at GWU and the fact that I had published that first refereed article, the PhD landed me a job in the English Department at Bowling Green State University—not my dream part of the country, but a good job at a time of a weak market and a job at an institution that had committed early on to an interdisciplinary approach to American Studies—where I soon began mining the dissertation for chapters I could revise and submit to scholarly journals. Without anyone to advise me on strategy, I probably went overboard on this, publishing four articles, or about three fourths of the dissertation, in refereed journals within my first three years at BGSU.[9] The danger inherent in this strategy was that when it came time for me to start turning the dissertation into a book, a university press might have argued that too much had already been published. I got lucky (not for the first time), and this did not happen.

When I did make the turn toward realizing every assistant professor's dream of a book, I knew I had to stick with the Nash model and bring my topic into the twentieth century. The adoption of the Adirondack Park State Land Master Plan (SLMP) in 1972 offered a neat parallel to Nash's concluding chapter on the federal Wilderness Act of 1964. The SLMP, moreover, transparently drew on the Wilderness Act for its definition of wilderness.[10] Addressing that and the twentieth-century context of the SLMP was my goal. I added three new chapters: one on how the state bureaucracy dealt with its wilderness for the first half of the century, another on how popular attitudes toward wilderness reflected the growing penetration of the science of ecology into the wider culture, and the third on the SLMP and how it represented an evolution of state policy away from an earlier resistance to wilderness protection to accommodation. All of these followed logically from the chapters I had already revised and all stuck with the Nash paradigm.

The completed manuscript was submitted to three university presses: Yale, because it was Yale and because it had published Nash's *Wilderness and the American Mind*; Syracuse (SUP), because it had a track record of publishing Adirondack books; and Temple, because it had a well-respected series of books in American Studies. SUP replied curtly and showed no interest. Yale's response was interesting. The editors sent me the two readers' reports, one of which, I was certain, had been written by Rod Nash. It said, though not in so many words, that the manuscript was pretty good, but just not Yale material. Temple sent the manuscript to the editor of its series in American Studies, Allan Davis, who responded with enthusiasm.

I had a publisher, and the book, titled *Forever Wild: Environmental Aesthetics and the Adirondack Forest Preserve*, appeared in spring 1985, with a cover sporting a somewhat doctored but evocative photo by Richard Linke and the title in a regrettable, almost unreadable font— just in time for my tenure review, which began in the fall of that year. Acutely aware of that crucial step in my academic career, I had chosen what later struck me as a pretentious, pedantic, and useless subtitle. "Environmental Aesthetics"—what does that even mean? And how does it help a prospective reader understand what the book is about?

Book reviews in academic journals are infamously and glacially slow to reach print, so I had to wait a long time before I had any idea of how the book was being received. Within about a year they began to materialize, and they were almost all positive. They surfaced too late to make any difference with BGSU's tenure committees, but that was of no consequence. One of the many absurdities of the academic gig is that at a mid-level university like mine in the 1980s it didn't matter whether or not your book was any good; it just mattered that you had a book. The line on your resume satisfied the dean, who didn't much care whether or not anyone was actually reading your book. Perhaps this has changed over the decades since 1986 when I was tenured and promoted to associate professor. But over the years, I could tell that people *were* reading my book. It was even used in a few college classes.

Of all the reviews *Forever Wild* received, the one that particularly registered with me was by Roderick Nash in the *American Historical Review*, the premier journal for historians practicing in the United States. One day I passed a colleague on the sidewalk at BGSU; he was coming out of the library and told me he had been browsing in the *AHR* and noticed a review of my book by Nash. Nearly quaking with anxiety, I ran to the journals section and began reading it.

Nash noted my "thorough research," but he criticized my lack of objectivity, by which he meant I was too invested in my own environmental values. He also liked the way I handled my primary sources: "The strength of Terrie's book is his textual analysis of these and other observers' accounts. No previous history goes into the literature with as much detail and perception." After noting that my book was a bit derivative, relying too much on the efforts of other scholars—including himself—Nash finished up with the kind of positive reinforcement a beginning scholar needs to hear: "What Terrie has done is to apply the broad cultural generalizations others have developed to a regional literature. The successful continuation of environmental history depends on this kind of refined focusing."[11]

In other words, my book was not particularly original, at least in its theoretical or methodological approach, and my obsession with wilderness was too obvious. Nash, of course, was exactly right. I had accomplished what I set out to do, which was to apply the model of mixing cultural and environmental history with which he had been so successful and see how it worked in the Adirondacks. I never claimed to be breaking new ground theoretically or methodologically or to be unbiased. I had a subject I loved to work with, and I did an adequate job with it. *Forever Wild* was not innovative, but I sure enjoyed writing it.

Neither Nash nor anyone else spotted a frightful and embarrassing error in *Forever Wild* (at least, no one ever mentioned it to me). In my chapter on the constitutional convention of 1894, I misquoted Article 7, Section 7, unforgivably dropping the words, "or be taken by any corporation, public or private" which precede the words, "nor shall the timber thereon be sold, removed or destroyed." The inclusion of these words in Article 7 was designed to prevent the deployment of eminent domain by either a state agency or a private entity aiming to use state land for any purpose other than what the delegates originally had in mind, hard as that was to identify precisely. My error was repeated in the front matter, where I opened the entire project with the crucial passage from the constitution.[12]

In 1994 Syracuse University Press, much to my delight, given its initial frosty response to the original manuscript, secured the rights to *Forever Wild* and issued a paperback reprint. A number of readers had caught typos and other mistakes but not this one, and I duly submitted a list of all I knew about, including this one, to SUP. To my horror, when the new paper edition reached me, the worst mistake of all remained uncorrected, as it does to this day. With this paperback I also seized the opportunity to deal with the issue of the subtitle, and we changed it to the more informative, though pedestrian, "A Cultural History of Wilderness in the Adirondacks."[13]

14

Long Lake

Just about the time I was finishing my dissertation, I got a call from Elliott Verner. He told me about an undeveloped lot on the northeast shore of Long Lake, adjacent to the property owned by him and Bill. It was for sale. The owner was W. Alfred Brim, an elderly bachelor lawyer living in Lockport, New York. I gave Mr. Brim a call.

I got right to the point: "Elliott Verner tells me that you have property on Long Lake for sale."

He replied, "I do."

"What do you want for it?" It so happened that he had asked a local realtor to take a look and tell him what was worth. He had just received the evaluation: $56,000. That was disappointingly not possible. So I said, "Well, thanks for talking to me, but that's way out of our range."

He immediately replied, "What can you offer?"

Where I came up with it I'll never know, but I said, "$20,000."

And without a moment's hesitation, he said, "Done."

We weren't finished. I then allowed as how I didn't have $20,000 and would have to look around to see where I could borrow it. And for the second time in this brief but already incredible call, he stunned me, a young man he had never met, with an instant reply: "Don't worry, I'll lend it to you." I later learned that Brim had inherited the lot—which had once been part of the vast holdings of the Santa Clara Lumber Company—from an uncle in the 1940s. He expected that someday he'd build a camp there. But he never had a family, and the camp never materialized. Once or twice he came to Long Lake and hired Herb Helms, Long Lake's bush pilot, to fly him down the lake to take a look at his parcel. On one of these visits he met Elliott. When he was in his

eighties and tidying up his affairs, he decided to sell the lot and called
Elliott to see if he knew anyone who might be interested; he wanted it
to go to someone whom the neighbors would approve of. Elliott mentioned it to a cousin, but the timing was off. He suggested it to a former student of his, but he passed. I was the third person Elliott called:
more good luck. Brim figured that if I was approved by the Verners
that was good enough for him. This was 1979, during the period of
inflation and interest rates in the high teens that made Jimmy Carter
a one-term president. Mr. Brim floated a three-year loan at 5.5%. And
he did the legal work, including filing the deed in Lake Pleasant, no
charge. More undeserved good luck.

Purchase was wrapped up by the spring of 1980. The lot was
just over twenty-six acres, with about seven hundred feet of rocky,
hemlock-lined lake shore—in high water, no beach at all, but in low
water, one roughly four-foot stretch of sand. Six miles by boat from
the Long Lake town dock. Elliott and his family were across a small
bay, with Bill and Abbie just the other side of them.

The property abuts the High Peaks Wilderness Area. I can walk
out my back door and into one of the largest publicly owned and
roadless areas east of the Mississippi. I can hike to Keene Valley or
Lake Placid, climbing Marcy along the way, without ever crossing a
road. I have easy access to my favorite Adirondack wilderness, Cold
River Country and the Sewards and Santanonis. As it is for many of
us blessed with the largely unearned privilege of owning one, camp
is part of my spiritual core. And my relationship with it, the town of
Long Lake, and all the wilderness surrounding it and me—from the
High Peaks to Lake Lila and the Whitney Wilderness—is a crucial
part of my evolving wilderness narrative.

The town of Long Lake is a fascinating, charming, occasionally
frustrating place, and although it looks much the same, it has changed .
in significant ways since I first laid eyes on it in 1966. The population has declined: from 900 in 1970 to 791 in 2020. Before 2020 it had
dropped even lower, reaching 711 in 2010. The school population is
less than half of what it was in the mid-twentieth century. The census
data do not tell us exactly how many school-age children are in a town,

but what they do show is telling: in 1970 Long Lake had 178 children aged five to fourteen; in 2020 it had only 53 in that age cohort. Conversely, in 1970, there were 273 people over fifty-five, while in 2020 there were 309.[1] The average age of a Long Laker is considerably older now than it was in 1970, and the number of families with young children is increasingly short of what the school district needs to run a viable school. There simply aren't enough kids for sports, drama, band, and most clubs. In the 2023–24 school year, there were fifty-nine kids, pre-kindergarten through twelfth grade, in Long Lake Central School.[2] A popular explanation for this demographic compression, in Long Lake and in similar towns in the Adirondacks (and around the country), is that the protection of wilderness in the Forest Preserve constricts economic development and thus means fewer decent jobs for young families with school-age children. If that were correct, my affection for the Town of Long Lake and its year-round residents would appear to run headlong into my wilderness obsession and my commitment to protecting it under all circumstances.

Hard times are nothing new for Long Lake. In the fall of 1933, with many Adirondackers suffering the ravages of the Great Depression, the Town of Long Lake was searching for a project to employ some of its destitute citizens. The town board voted to spend $3,000 to create a pond by flooding about thirty-five acres of wetland at the outlet of Shaw Brook. An engineer's report predicted that the work, employing sixty men for sixty days, "would go far toward the relief of unemployment." The project was designed to coincide with and take advantage of the state's plan to improve the highway from Long Lake to Tupper Lake. The engineer submitted his report to the town board in the late fall, recommending that work begin as soon as the weather improved the following spring. But the board declared that work would commence immediately: there were families in town experiencing genuine hunger, and the board was determined to put a paycheck in the hands of many men as possible as quickly as possible.

Before this project was proposed, the highway passed via a low bridge from the site of the Adirondack Hotel to Pine Island, at the south end of the current bridge. In 1933–34, the state built a causeway from the hotel to Pine Island. Simultaneously, two causeways were constructed by the town. One, of about 640 feet, linked the west shore, at the site of the Sagamore Hotel, to an unnamed high spot, now an island, near the outlet of Shaw Brook. The other, about 560 feet, ran from there to Pine Island, with a spillway over which the waters of the new pond dropped into Long Lake (as they did until July 11, 2023, when the causeway was breached after a cloudburst that flooded Shaw Brook). The new pond and the park beside it, finished in November 1934, were named for Town Supervisor Lewis L. Jennings, who had died in October 1933.[3]

The story of the construction of Jennings Pond and Park is more than a satisfying tale of small-town resilience—though that is at the core. Much of the money that passed to the destitute came from outside the town, in the form of taxes paid by nonresident landowners and by the State on its lands in the Forest Preserve, all protected as wilderness by the constitution. The story of a small town creating work for its needy citizens is complicated. It's also an illustration of how outside money is essential to keeping such towns afloat. Year-round residents of the Adirondacks like to think of themselves as self-reliant and independent, and in so many ways they are. But it's important to note the web of support that flows into the towns of the Park from downstate. The budgets of the Long Lake Central School District, the town of Long Lake, and Hamilton County depend on tax payments on those wilderness lands (in addition to the money spent in town by tourists), none of which require school buses, road maintenance, or all the other amenities and infrastructure paid for by local government.

Long Lake Central School, right up the hill from Jennings Pond, is a locus of the town's cultural life and a major feature of town identity, and it illustrates this relationship. It's a tiny school and could not possibly be maintained without tax revenues coming from outside sources: the state of New York, which pays taxes on the roughly fifty percent of the town that's in the Forest Preserve, and seasonal camp owners, who do not send their children to LLCS.[4] The small student populations of several central Adirondack school districts periodically inspire talk of consolidation, but the idea of merging with a neighboring district is anathema to most Long Lakers. There is considerable logic to the establishment of a new school district serving Indian Lake (with which Long Lake now shares team sports), Blue Mountain Lake, Long Lake, and Raquette Lake. But there is virtually no enthusiasm for this in any of the towns. If the State were not paying taxes on the Forest Preserve and its constitutionally protected wilderness, sentiment on this subject might be different.

As nearly every demographer will point out, the good-paying jobs for young adults—precisely the population cohort a town needs to keep its school viable—are in or near metropolitan areas. This means that the graduates of Long Lake Central School who go off to college—as many do—are unlikely to come back home to start a family. It's a trend one can find throughout rural America. The growing population of retirees in Long Lake is good for the local economy, but it doesn't help the school. The retirees have a rich life, with group hikes, kayak paddles, ski expeditions, and well-attended pot-luck dinners. But they do not have kids in a K–12 school.[5]

In 2014, a widely circulating pamphlet published by the Association of Adirondack Towns and Villages (known as the APRAP Report) explicitly insisted that conservation of open space, including wilderness designation, restricted economic opportunity for young families and consequently led to declining numbers of children in Adirondack schools.[6] A local example of the same local conviction occurred beginning in 1973 when Long Lake's Herb Helms, accustomed to flying hunters and anglers into the backcountry, sued the State after the original State Land Master Plan, signed by Governor Nelson Rockefeller in 1972, listed Round Pond and other remote water bodies in Wilderness Areas; motors, including float planes, are prohibited in designated wilderness.[7] Helms claimed that the State was depriving him of income, and he was supported in this claim by local government. He lost the suit, but the certainty that classifying state land as wilderness limits economic opportunity for local businesses persists, both in Long Lake and throughout the Park.

In 1998, the State bought the stunning Little Tupper Lake from Whitney Industries and shortly thereafter created the William C.

Whitney Wilderness Area. At the time the State acquired it, Little Tupper hosted a rare strain of native brook trout. Within a few years, bass appeared in the lake, and the trout population collapsed.[8] A rumor circulated at the time that the bass had been intentionally introduced by someone offended by the wilderness designation and certain that it was contrary to local interests.[9]

In 2019 Protect the Adirondacks! demolished the frequently repeated claim that conservation is bad for the local economy.[10] I worked closely with Peter Bauer and James McMartin Long, who produced a thorough examination of local demographics and economics. We looked at rural populations in New York State, in the northeastern United States, and across the country. We found no evidence that the economic or demographic realities in the Adirondacks are any worse than those in similar rural regions, including many with minimal acreage in any form of conservation, throughout the Unites States. Indeed, the Adirondack economy is doing better than that of many parts of rural America, and this can reasonably be attributed to the appeal of protected open space, including designated wilderness.

This is not to say that life is not challenging in the Adirondacks. The assessment of local circumstances in the APRAP report is in most respects accurate. It found an aging population, a shortage of young adults to serve as first responders in volunteer fire departments and rescue squads, and declining school populations. We at Protect had no quarrel with these diagnoses and the identification of Adirondack problems that are shared throughout rural America. Rural America is in trouble. It is not surprising that when I was recently talking to a Long Lake businessman and the subject of the Park Agency and protected wilderness came up, he repeated an often-heard local talking point, that the Park Agency, environmentalists in general, and especially those of us putatively obsessed with wilderness, "want us all out of here." The notion that environmentalists somehow aim to remove completely the Adirondack year-round population in order to create a vast wilderness—a playground for downstate hikers and bird watchers—pops up routinely in regional blogs and bar-room conversations. It is, in fact, hard to run a reasonably profitable small business anywhere, and it's especially hard in a place like the Adirondacks where so much depends on income generated during the warm months. I don't blame my Long Lake friend or anyone else for feeling beleaguered, overworked, and underpaid. I do blame those behind the APRAP report and others for promoting mythology for which there is no evidence.

All this makes the topic of wilderness in the Adirondacks confusing and ambiguous. To some of us wilderness, however we define it, is sacred, something to be defended and treasured. To others it is ridiculous, an impediment to the good life and the honest pursuit of the American Dream. For them it's a fiction manufactured by the vague yearnings of affluent outsiders who have a demonstrably more powerful voice in Albany than do the 130,000 year-round residents of the Park.

15

William Cronon

It's unusual for professors of environmental history (or of anything else) to publish in the *New York Times Magazine*. But on August 13, 1995, William Cronon, the Frederick Jackson Turner Professor of American History and Geography at the University of Wisconsin and one of the leading figures in the rapidly growing field of environmental history, did just that, and it was a bombshell.[1] His title, "The Trouble with Wilderness," was a deliberate echo of Robert Marshall's path-breaking article of 1930 and an explicit announcement that he intended to take a hard, perhaps skeptical look at Marshall's argument about the need to preserve American wilderness.

Cronon opened with an acknowledgment that wilderness has been a major focus of the American environmental movement, largely because wilderness to so many people represents the antidote to a culture increasingly corrupted by urbanism, industrialism, and consumerism. So far, so good; no toes stepped on. But he quickly pivoted and began a lengthy and insightful critique of a facile reliance on unexamined assumptions. "Far from being the one place on earth that stands apart from humanity, it [wilderness] is quite profoundly a human creation—indeed, the creation of very particular human cultures at very particular moments in human history." In other words, he argued, wilderness is not so much a geographical or ecological reality as it is a projection of cultural needs that are tightly and inextricably connected to their historical moment. Cronon does not dispute the reality or the manifest appeal of the rocks and trees and water that we encounter in the backcountry. What he questions is how these add up to the abstraction that we call wilderness. The very fact that what wilderness meant to

William Bradford in 1620 is so different from what it meant to many Americans in the late twentieth century when he wrote this essay reinforces his point that wilderness is a cultural construction.

Cronon outlines two familiar explanations for how many Americans pivoted from Bradford's sense of wilderness as terrifying to John Muir's sense of the wilderness as sacred. One was Romanticism's emotional conviction that wild scenery and wild places were spiritually inspiring. The second was Americans' fixation on their own mythologized frontier past and the apprehension that if wilderness was not preserved, the locus where ostensibly American values like self-reliant individualism and manly virtue were forged would vanish and those very values would erode or even disappear. Robert Marshall, I'm sure, would have nodded approvingly. These were fundamental to his argument for why wilderness needed protection.

After retracing these familiar themes, Cronon moved to his critique, beginning by pointing out the elitist and sexist assumptions behind the environmental politics of such early advocates of the wilderness mystique as Theodore Roosevelt and Owen Wister. To them and their brethren (few devotees of wilderness in their day were women, largely denied by their cultural circumstances the opportunity to even experience the American wilderness), the wilderness was a place for masculine men of a certain class.[2]

In addition to the sexism and elitism, moreover, was the racism. The confluence of the wilderness and frontier themes depended on the idea of the wilderness as outside the world of human imprint, uninhabited, pristine, and unaltered by the corrupting impact of any human cultures. The manifest existence of Indigenous peoples, for whom what affluent White men called wilderness was the place where they lived, raised their children, and went about the daily routines of life, was erased. Worse, many of the places most sacred to the late nineteenth-century cult of wilderness had seen their Indigenous inhabitants forcibly removed solely so that those well-educated and largely affluent White men could pretend they were Daniel Boone as they hunted, fished, fraternized, and smoked cigars amid the splendors of undeniably sublime scenery. Yellowstone, Yosemite,

Glacier—among the earliest of the western locales set aside as tourist and camping destinations—had been the ancestral homes of Native peoples.[3] Cronon notes, "The removal of Indians to create an 'uninhabited wilderness'—uninhabited as never before in the human history of the place—reminds us just how invented, just how constructed, the American wilderness really is."

The erasure of centuries of Indigenous presence on these lands led to what Cronon argues is an erasure of history itself, notwithstanding the inclination of Marshall (and others) to fantasize about how wilderness explicitly puts him *in* history. When we insist on an illusory absence of human occupation and impact, we set up an escape from the human past, creating a place where we pretend history has not happened, a place that is timeless and immutable. We are understandably searching for a place not subject to the putatively sinister forces of modernity, and we project this need onto a few special places that we call wilderness. Moreover, Cronon argues, those people—e.g., Headley, Roosevelt, Marshall, myself—most eager to lose themselves in these constructed idyllic retreats are likely to be educated and comfortable individuals whose lives have not been defined by the sort of actual physical labor and hardship that defined the actual American frontier: "The dream of an unworked natural landscape is very much the fantasy of people who have never themselves had to work the land to make a living—urban folk for whom food comes from a supermarket or a restaurant instead of a field, and for whom the wooden houses in which they live and work apparently have no meaningful connection to the forests in which trees grow and die."

At this point I think Cronon's rhetoric is getting a bit out of hand. It is possible, for example, for someone to find spiritual sustenance in a designated wilderness area, where logging is prohibited, and still be perfectly aware that both two-by-fours and paper depend on logging and the physical labor of the loggers who harvest trees and send them to market. But the point about the class origins of so many mid-twentieth-century wilderness advocates and the social distance between them and, say, loggers and farmers, is compelling. It is neatly expanded by Richard White in an essay in the same volume in which

Cronon's "Trouble with Wilderness" appears, "'Are You an Environ-mentalist, or Do You Work for a Living?': Work and Nature."[4]

The fetishization of wilderness, notes Cronon, leads to a classic form of dualism, a sense that wilderness is the exoticized other and that human beings, ourselves, are apart from nature rather than being a part of it. "Any way of looking at nature that encourages us to believe we are separate from nature—as wilderness tends to do—is likely to reinforce environmentally irresponsible behavior." And this in turn leads to a failure to acknowledge our responsibility to nature not blessed with a wilderness designation: "Idealizing a distant wilderness too often means not idealizing the environment in which we actually live, the landscape that for better or worse we call home." We ignore, argues Cronon, the pond at the city park or the wood lot on the next street.

If one of the chief appeals of wilderness to those of us in its cult is the opportunity for spirituality, Cronon argues, we miss opportuni-ties for spiritual enrichment closer to home. "The tree in the garden is in reality no less other, no less worthy of our wonder and respect, than the tree in an ancient forest that has never known an ax or a saw—even though the tree in the forest reflects a more intricate web of ecological relationships." He's right, of course. The tree in the backyard or the tree at the curb outside our urban apartment is indeed worthy of our wonder. We should all welcome the opportunity to become entranced by a glimpse of so common a bird as a cardinal or a goldfinch outside the dining-room window. Where I find Cronon unconvincing is his insistence that dwelling on the richness of the wilderness experience limits our awareness of such opportunities closer to home. It was my good fortune to have extensive wood lots, often vaguely domesticated, to play in when I was a kid, and those experiences led to and were a foundation of my love for the Adirondack backcountry. I believe Cronon is wrong when he argues, "Without our quite realizing it, wilderness tends to privilege some parts of nature at the expense of others." I can think of no wilderness defender for whom this was ever the case. Not even John Muir.

Beyond the points where I quibbled with certain details of Cronon's argument, it is undeniable that he touched a raw nerve. I was building my academic career and my spiritual life on a certainty that wilderness existed and that it was an altogether good thing. The possibility that it was a cultural construction, resting on a foundation of elitism, sexism, and racism, and that a fixation on wilderness could lead to environmental irresponsibility stung. I well remember the August morning in 1995 when I read the initial version of this essay in the *New York Times Magazine*. I picked up the *Times* at Northern Borne, Long Lake's invaluable grocery store, and took it back to camp for what was the usual Sunday ritual of too much coffee and reading the paper on the deck. I looked carefully at the easy places where I could disagree with Cronon—mainly the argument about how a wilderness obsession can distract us from other environmental issues—and ignored the rest.

But I could not deny for long that Cronon was on to something important. It wasn't the first time I had encountered the thesis that wilderness was largely a projection of American, male values onto physical nature; this had become a routine declaration at some of the scholarly meetings I occasionally attended, mainly the American Society for Environmental History and the Western Literature Association. But Cronon made the most forceful case I had seen, and he presented it outside the academic cloister.

The idea that our sense of the world around us is socially constructed had been a powerful force in American Studies for many years, receiving its most influential push from a brilliant and innovative book, *The Social Construction of Reality: A Treatise in the Sociology of Knowledge*, published in 1966 by sociologists Peter L. Berger and Thomas Luckmann. To reduce a complicated book to its useful essence, Berger and Luckmann argued, among many other things, that our understandings of reality reflect institutionalized and commonly shared values. What we think we know, what we think is merely common sense, profoundly affects our perceptions of everything. In other words, there is in fact a physical reality out there, but our understandings of it are shaped by our personal history, our shared experiences, and what we absorb often unconsciously from the values of the culture around us.

But the realization that our cultural and institutional understandings are socially constructed can be misleading. My first reading of Cronon suggested that he was insisting that wilderness is a lie, a deception, that there is no such thing as wilderness. He seemed to be saying that this idea of wilderness with the concomitant need to protect it was a trick played on everyone else by rich White men. My reaction was a combination of anger and dismissal. Behind that was some vestigial and unexamined guilt for subscribing to such an elitist and racially charged enterprise.

In the meantime, Cronon found himself the object of a furious counterattack, from Alexander Cockburn at the *Nation* among others. In his column at one of the country's most widely circulating leftist magazines, Cockburn, shooting eloquently from the hip, as he nearly always did, blasted Cronon for undercutting wilderness protection, noted that he was the editor of a series of scholarly books on environmental history underwritten by the ethically suspect Weyerhaeuser Corporation, and insisted that Cronon's recent election to the governing board of the Wilderness Society was proof that the country's oldest environmental lobby devoted exclusively to wilderness protection had abandoned its core mission. Cronon responded that Weyerhaeuser had no way to influence his editorial decisions and that Cockburn's assertion that wilderness "is merely a state of mind . . . grossly distorts my position."[5]

That Cronon had become a trustee of the Wilderness Society (founded, we should recall, by Robert Marshall, among others) is telling. So is Cockburn's apparent failure to understand the concept of cultural construction. Cockburn missed much of the meat of Cronon's argument. Cronon explicitly explained that he supported "efforts to set aside large tracts of wild land." This explains, we can assume, the willingness of the Wilderness Society to invite him onto its governing board.

Cronon was no enemy, it turned out, of managing some parts of our landscape in certain ways and identifying them as wilderness—precisely what we do in the federal lands protected by the 1964 Wilderness Act and in those parts of the Adirondack Forest Preserve set aside as Wilderness Areas in the 1972 State Land Master Plan. But he strenuously objected to facile assumptions whereby White kids like me—or even grizzled wilderness protectors like Edward Abbey or Dave Foreman—could pretend they were camping in places on which no human impact had ever occurred. In an article on "The Pristine Myth," geographer William Denevan (a colleague of Cronon's at the University of Wisconsin) quoted a well-known saying attributed to Chief Luther Standing Bear of the Oglala Sioux: "Only to the White man was nature a 'wilderness.'"[6]

As I talked to other environmental historians and pondered the implications of Cronon's thesis, it was inevitable that my thinking would evolve. It was easy to dismiss the argument that emphasizing wilderness protection led to indifference to other environmental concerns, especially those closer to home. Indeed, it could be argued that this point was exactly backward and that many an activist on the front for cleaner water and air in our cities or for environmental justice in general had first caught the environmental bug on a camping trip in the backcountry. That was certainly the case with me.

But the larger case, starting with the elitism and racism of so many of the earliest wilderness advocates, was hard to contest. I had known for years, in fact, that Joel T. Headley, whose 1849 book, *The Adirondack: or, Life in the Woods*, was one of the major sources for a chapter in my *Forever Wild*, had been an apologist for American imperialism in the Mexican War and that he had published a xenophobic screed attacking immigration, especially Irish immigration, to New York City.[7] Beyond the Adirondacks, recent scholarship has cast a new light on such previously sainted figures as John James Audubon and John Muir, whose racial views have been dissected and challenged.[8]

There was no bingo, slap-the-forehead moment, but what I ended up with was the understanding that wilderness is of course a cultural construction. And much of its history, or at least the history of some of its evangelists, is not pretty. But what matters is to see that even if it is a fiction, something can be a cultural construction and also be a compelling reality. This is something that many people who have read Cronon's essay miss.[9] They digest the many altogether sensible things he has to say about wilderness and then conclude that they, or, more likely, others, have been deluded.

In any culture so much of what is truly important is culturally constructed, and there is no better example of this in American culture than race. We now know, or at least we should know, since scientists and geneticists who actually study such matters have told us so clearly, that race is an invention. It has no more biological significance than the color of our eyes. Much more important, it means nothing when we consider such human characteristics as intelligence, generosity, honesty, or dignity or, for that matter, brutality, cruelty, deceptiveness, or ignorance—or any of the other human features that we inevitably ponder when we think about what distinguishes one person from another.

Yet is there anything more important in American history than race? Race is *the* American story, from a beginning that included erasure of and theft from Indigenous, non-European people and an economy built on chattel slavery, through a Civil War fought because of race-based slavery followed by an era of Jim Crow and widespread lynching, to the present when entire states are aggressively trying to prohibit the teaching of the facts of American racial history. Race in America embraces so much of our narrative, both brutality and compassion, both reactionary violence and progress, both White supremacy and the certainty among at least some Americans that all people truly are created equal. Yet, as any serious anthropologist will tell you, in a biological sense there is no such thing as race.

All of which is to say that wilderness is both a social construction and something demonstrably real, with a powerful presence and deep personal meaning in the lives of many Americans. With great insight and precision historian Paul Sutter has discussed this endlessly provocative feature of the wilderness conundrum. Is it real? How do we continue to defend it once we ponder the "social, ethnic, and racial biases" of so many of its advocates? Can we even discuss such a thing? It has become too easy, writes Sutter, for the nonbelievers to dismiss the wilderness idea as "ecologically naive, dispossessive, class biased, consumerist, and hopelessly separated from concerns for social justice." Such a dismissal tends "to caricature the wilderness idea and wilderness advocates." Sutter goes on to offer a meticulous study of the founding of the Wilderness Society and to show how Robert Marshall, Aldo Leopold, Benton MacKay, and others developed sophisticated arguments for why wilderness is indeed an important asset—ecologically, spiritually, philosophically—for human societies.[10] In this book I am trying to develop those same themes.

16

Wilderness and American Studies

Cronon's provocative intervention in wilderness debates coincided with a reexamination of one of the key texts of both American Studies and environmental history. *Wilderness and the American Mind* offers a classic example of the methodology of the myth-symbol school, inescapably signaled by the title, which claims to assess "the American Mind." Like other memorable works of that era, e.g., Smith's *Virgin Land*, Leo Marx's *The Machine in the Garden*, and John William Ward's *Andrew Jackson: Symbol for an Age*, to name a few of the most prominent titles, *Wilderness and the American Mind* traces the evolution of an idea in what scholars of that era were confident was the unified and identifiable American mind.[1] Like Smith's *Virgin Land*, Nash's book draws on canonical and popular literature, government documents, and a wide array of other sources, and deftly interweaves textual analysis with how an evolving narrative eventually influenced policy.

It's a magnificent book; in its heyday it doubtless nudged many a graduate student, including me, into the field of environmental history, of which it is an early classic. Roderick Nash was himself an enthusiastic participant in the back-to-nature movement of the 1960s, and in this book he managed perfectly to wed his scholarly and spiritual impulses, realizing a goal that has frequently characterized scholars in both environmental history and ecocriticism. Nash was my model of what an academic could be.

But *Wilderness and the American Mind* also manifests the flaws we have come to associate with the myth-symbol school.[2] With its initial insistence that there actually is such a thing as an "American mind" and its reliance on the published writings of literate, mostly wealthy

White men, it homogenizes and whitewashes American culture. Native cultures appear, of course—inevitably, one supposes, in a work on this topic—but never as active players in the drama of how understandings of wilderness evolved. The book appeared long before Annette Kolodny showed us how intensely gendered masculine accounts of the wilderness inevitably are.[3] And I don't think there is a single African American account mentioned anywhere. I don't mean to dismiss the book on account of these omissions. During most of the time Nash was examining, it was elite White men who made land-use decisions, and his primary focus was on how popular attitudes eventually led to policy. But these days, no one would try to assess anything so huge and amorphous as the "American mind"—or even claim that such a thing exists—without an explicit acknowledgment of American diversity and how marginalized voices are so routinely erased from both our history and the processes by which we reach land-use policy decisions.

In addition to underplaying gender and race as driving forces in American culture, *Wilderness and the American Mind* also ignores class—one of Cronon's key issues. Throughout the twentieth century, when battles over wilderness protection developed in all their fury, profoundly different class understandings of what nature is and what it is good for were prominent in those struggles and have been ever since. As Cronon was one of the first to insist, we cannot adequately discuss the story of wilderness in America without recognizing the role of class in defining, eliminating, protecting, and resisting it.[4]

Wilderness and the American Mind is a declaration that wilderness exists, that the story of wilderness in American culture is a slow but relentless and apparently inevitable embrace of it by ever-growing numbers of Americans, and that this inchoate American acceptance of wilderness constitutes a triumph of our culture. In other words, it's a Whiggish tale of the past leading inexorably to us, with "us" being rather narrowly defined.

In my faithful, largely unquestioning application of the Nash model to the specific case of the Adirondacks, I reproduced the major flaws of his method, both in my dissertation and in *Forever Wild*.

My primary sources were almost exclusively by educated and comfortably well-off White men. Where Nash had a Concord intellectual like Henry David Thoreau, I had his contemporary, the popular historian and travel writer Joel T. Headley.[5] Where he had Western explorers like John Wesley Powell, I had Powell's contemporary, Adirondack surveyor Verplanck Colvin (who I believe was profoundly influenced by what he read about the great Western surveys and was unconsciously searching for a way to recreate their adventures in his beloved Adirondacks). Where Nash benchmarked the establishment of the first National Park, Yellowstone in 1872, and the 1964 National Wilderness Act, I used the conservation measures adopted by the state of New York to protect the Adirondacks (and Catskills) in the 1880s and '90s and then the establishment of the Park Agency and its State Land Master Plan in the 1970s, nearly all the work (with a few notable exceptions) of comfortably well-off, educated White men.

What did women think about the Forest Preserve in 1894? What about the recent immigrants toiling in New York's fast-growing industrial sector and living in the fetid tenements of New York's crowded and polluted cities? What about the descendants of the Iroquoian and Algonquian peoples who had hunted and lived in the Adirondacks for centuries? I don't know whether I could have answered these questions, but I do know that trying to track down a broader base for my conclusions did not occur to me.

I noted with enthusiasm, for example, in both the dissertation and *Forever Wild* that when Thomas Jefferson visited Lake George as a tourist in 1791, he described his rapturous response to the scenery in the Burkean rhetoric of the sublime and beautiful; it was a response characteristic of a literate, well-read Euro-American intellectual.[6] What did not occur to me was to consider that Jefferson was accompanied by James Hemings, his Black servant, an enslaved Virginian, brother of Jefferson's also-enslaved mistress, Sally Hemings. I did not even know that James Hemings was there until I read Annette Gordon-Reed's magisterial *The Hemingses of Monticello*.[7] We cannot know what Hemings thought of the wilderness surrounding Lake George, but failing to acknowledge his existence and the likely possibility that he was just as moved by it as Jefferson was illustrates a classic failure of the myth-symbol method.

Likewise, as Daegan Miller has eloquently suggested, we should ponder what Black settlers, given land by philanthropist Gerrit Smith and encouraged to homestead in the Adirondacks before the Civil War, thought of the dense and apparently pathless forests in which they found themselves and where they tried to carve out farms. We can be reasonably sure their understanding of the Adirondack wilderness was profoundly different from that of their contemporary, Joel T. Headley, who described his Adirondack wilderness adventures with glowing, Wordsworthian warmth. Learning of the plans for this experiment, the abolitionist Frederick Douglass, a Black man and a friend of Smith's, wrote in his newspaper, *North Star,* "Advantage should be at once taken of this generous and magnificent offer. . . . The sharp axe of the sable-armed pioneer should be at once uplifted over the soil of Franklin and Essex counties, and the noise of falling trees proclaim the glorious dawn of civilization."[8] To Black families contemplating a pastoral life in the Adirondacks, wilderness was exactly what it was to other poor Americans heading at the same time to America's Midwest and beyond: something to be eliminated as quickly as possible before prosperity or even survival could be secured. Wilderness as a locus of spiritual or physical redemption was an affluent White man's domain.

If the primary sources existed, an equally interesting contrast to Headley's enthusiasm might be found in how Mitchell Sabattis, an Abenaki who lived in Long Lake and who served as one of Headley's guides on his several lengthy camping trips, responded to the wilderness where he made his living. Sabattis became a legendary figure in the mid-nineteenth-century phenomenon of "sportsmen" like Headley who constructed the popular narrative of the Adirondack wilderness as a seasonal recreational retreat for professionals from Eastern cities. Melissa Otis has argued persuasively that Sabattis and other guides— both Indigenous and White—adopted a carefully constructed persona to fit the culturally determined stereotypes that Headley and his ilk brought to their encounters. Both Sabattis and well-known White guides like Harvey Moody of Saranac Lake understood that what was wilderness to Headley was the place where they labored for their clients and where they enacted a delicate performance of rusticity and woodcraft to suit urban expectations.[9] To Mitchell Sabattis and many others, the wilderness was a dramatically different place from the wilderness constructed by the "American Mind" as understood by Roderick Nash.

17

Contested Terrain

Throughout the years I taught at BGSU, I maintained close ties to the Adirondack Museum. I made sure to establish cordial connections with each successive director, and I continued to use the library for research, where I was generously indulged, especially by Jerry Pepper, who reigned serenely over and expanded the museum's collections for decades. Sometime in the early 1990s I was talking to Director Jackie Day. She was lamenting the absence of a short pamphlet-length history of the Adirondacks, and I leaped at the opportunity to write one. I had a number of goals here. First, simply to enjoy the opportunity of putting together a condensed history of my favorite subject. And second, to develop a correction to *Forever Wild*, based on my reading of Cronon's influential essay. That book's reliance on elite sources and its failure to deal with social history and the importance of the year-round residents continued to bug me. Needless to say, these goals quickly enlarged the project, and I was writing another book.

Before I explicitly acknowledged it, what I was doing was in fact *rewriting* a book I had already written, arguing with myself. This was not the way to establish a reputation as a serious historian. Any other youngish historian would be looking for a topic quite distinct from the first book—in the same general field, of course, such as environmental or social or cultural history, but not as close to a previous book as my new project was to *Forever Wild*. I was painting myself into a provincial corner, but I didn't much care. I was tenured at BGSU and had been promoted to full professor in 1992, on the basis of some scholarly articles on non-Adirondack topics in solid journals. And, almost certainly, because of an ever-tightening academic labor market, I was unlikely ever to score what I really wanted, a lateral move to a college in New York. I had the luxury to work on what I wanted to work on rather than on what might improve my reputation among other environmental historians. I was quite willing to forgo trying to be an historian with breadth and to be known for, I sincerely hoped, sticking to and doing a decent job on a narrow subject.

Contested Terrain: A New History of Nature and People in the Adirondacks was jointly published by the Adirondack Museum and Syracuse University Press in 1997. I returned to many of the same sources I used for my first book, but I looked at them, particularly as they reflected attitudes toward wilderness, from a new perspective. Now, influenced both by myth-symbol revisionism and by Cronon, while certainly continuing to take Joel Headley, for example, seriously and appreciating the genuineness of his appreciation of the Adirondack wilderness, I also saw Headley as a representative of an elite, professional class eager to appropriate the Adirondacks for its own cultural purposes. In other words, Headley and his ilk constructed a wilderness to suit their cultural needs, just as Cronon argued that Theodore Roosevelt and others had done elsewhere. At the same time, Headley and writers like him simultaneously condescended to and romanticized the year-round residents who guided their patrons and kept them alive. Their voices—that is, the voices of guides like Mitchell Sabattis and Alvah Dunning or anyone of their class—were conspicuously absent from *Forever Wild*, just as the voices of similar working-class people were absent from Roderick Nash's *Wilderness and the American Mind*.

Contested Terrain was well received and reviewed. It is still in print, with an additional chapter added in 2008. It is often used in college classes, and I'm delighted that people appear to read it. But so far as I know, none of its readers, except two, have noted just how it relates to *Forever Wild*. Both of those readers understood how *Contested Terrain* was a corrective, but they both also suggested, inaccurately I believe, that it was a repudiation of the earlier book. The first was Jim German, professor of history at SUNY-Potsdam, in an essay surveying and assessing various books on Adirondack history.[1] The second was Daegan Miller, a PhD student in history at Cornell whom I met shortly after I moved to Ithaca in 2012. He was at work on a dissertation in environmental history with an innovative chapter on the Adirondacks and the antebellum experiment in African American settlement around North Elba and southern Franklin County. There he referred to my two Adirondack books and in a lengthy endnote perceptively noted how one led to the other. He also identified the influence of William Cronon:

> His [i.e., my] *Forever Wild: A Cultural History of Wilderness in the Adirondacks* (1994) is very clearly written in a Roderick Frazier Nash, triumph of wilderness, cultural and intellectual history vein, which makes his *Contested Terrain: A History of Nature and People in the Adirondacks* (1997) all the more remarkable, for *Contested Terrain* is a complete revision—one might almost call it a disavowal—of *Forever Wild*. It's much more of a social history, written according to the new, post–"The Trouble with Wilderness" paradigm, and insists on a declensionist story in which the triumph of wilderness has meant the erasure of the "real" Adirondacks.[2]

Miller is correct to note the shadow cast by Cronon over *Contested Terrain*, but he is stretching the case to see it as a "complete revision." My goal was not to cast the protection of Adirondack wilderness as a "declension," or a step in the wrong direction. To the contrary, I believed then, and believe now, that wilderness protection was and is a positive movement. My point was to show how year-round residents of the Adirondacks saw things differently and how their views on the matter have, until relatively recently, been consistently ignored. It wasn't that *Forever Wild* was wrong; it was incomplete. In the same way, I think, we should see that Henry Nash Smith's *Virgin Land* and Leo Marx's *The Machine in the Garden*, to name two of the best-known examples of the myth-symbol school, are not so much wrong as incomplete. *Contested Terrain* added what was missing from *Forever Wild*. Annette Kolodny's *The Lay of the Land* and *The Land Before Her* added, brilliantly, what was missing from *Virgin Land*.

The primary device in developing my argument in *Contested Terrain* was the idea that when we think about the land, we tell stories about it. We develop narratives, coherent (mostly, but often not) collections of beliefs and assumptions about the land's meaning and best use. My deployment of this argument was an obvious reflection of my understanding of how our perceptions of the world around us are socially constructed. Probably the best example of this device in *Contested Terrain* is the comparison of how Joel T. Headley and John Todd characterized the settlement and environs of antebellum Long Lake. They were talking about the same place yet described it profoundly differently. To Headley, Long Lake was a paradise for hunting and fishing, an ideal destination for urban professional men like him, where they could restore body and soul amid the healing glories of the American wilderness. To Todd, Long Lake was a frontier settlement, where sturdy but poor American families were scratching out a living and pursuing the Jeffersonian dream of the middle landscape, that mythic land between the wilderness and the city where wholesome yeomen and -women, uncorrupted by urban woes, could till the soil and promote the true essence of American democracy.[3]

That Headley's vision prevailed and Todd's was largely forgotten at the time the State instituted protections in the 1880s and '90s seemed to me to demonstrate forcefully the validity of Cronon's argument about the elite nature of wilderness protection. Headley's notion of wilderness, projecting the needs of elite hunters and anglers, led to critical developments in land-use policy. But, as I hope I made clear, I saw this not as a step in the wrong direction, as Miller concluded that I did, but as an unintended but fortunate consequence. Headley, to be sure, did not adequately acknowledge the lives, fortunes, and understandings of local men like Mitchell Sabattis or John Cheney, both of whom served him as guides and kept him fed and comfortable, but Headley did recognize something of great value in the Adirondacks. And that was the persistent Adirondack wilderness.

18

The Constitution, 1915

In the years after the first edition of *Contested Terrain*, with its emphasis on cultural history and the construction of popular narratives, I began to realize that I had paid inadequate attention to the twentieth-century constitutional and judicial context of wilderness in the Adirondacks. Yet again, I started reworking what seemed to be my only subject. I didn't know it yet, but I was preparing the ground for my participation in *Protect*.

In 1914 New York voters were asked whether they wanted a convention to consider a new constitution; the voters said Yes, and a convention began its deliberations in June 1915.[1] On August 6, 1915, Delegate Dow, from Chautauqua County, representing the Committee on Conservation, introduced the report of his committee, which had considered all proposed modifications of Article 7, Section 7, of the 1894 constitution.[2] He outlined the importance to the state of the canals and asserted that to protect "the living waters and their economic flow . . . the forests . . . shall be preserved inviolate forever." He asserted that the 1894 convention, by adding protection for the Forest Preserve to the constitution and thus removing the legislature from the process, was following the precedent of what the state had done with canal lands, thus reemphasizing the importance of an intact Forest Preserve to the state's commercial system. He noted that an amendment passed by legislature in 1913 and 1915 (two passages required to get an amendment on the ballot) to initiate a constitutional amendment to permit limited logging on the Forest Preserve supplied the primary context for his committee's deliberations.[3]

The convention of 1915 thus faced an immediate threat to the protection provided by Article 7, just as the convention of 1894 contemplated the potential damage looming in the legislation of 1893 to encourage logging on the Forest Preserve. Dow declared that this move on the part of the legislature was a terrible error and had not been submitted to the people for a vote only because of a procedural error. An important element of his committee's report was a recommendation for the creation of a nine-member commission, to be called the Department of Conservation, to manage the Forest Preserve.[4]

In (for our purposes) the critical finding of the Committee on Conservation, Dow noted that it was recommending a significant change in what had been Article 7, Section 7:

> We have retained the language of the present Constitution, adding the words, "trees and," for the purpose of making more inclusive the scope of the provision; and, to obviate some of the narrow constructions which have been placed upon the present Constitution, we have added that the commission is, however, empowered to reforest the lands in the Forest Preserve, to construct fire trails thereon and to remove dead trees and dead timber therefrom for the purposes of reforestation and fire protection solely, but shall not sell the same.[5]

The language of the proposed article, which would be Section 2 of a revised Article 7, read thus:

> The lands of the State, now owned or hereafter acquired, constituting the forest preserve as now fixed by law, shall be forever kept as wild forest lands. They shall not be leased, sold or exchanged, or be taken by any corporation, public or private, nor shall the trees and timber thereon be sold, removed or destroyed. The department is, however, empowered to reforest the lands in the Forest Preserve, to construct fire trails thereon and to remove dead trees and dead timber therefrom for the purposes of reforestation and fire protection solely, but shall not sell the same.[6]

This was followed by one further sentence about the construction of a highway from Saranac Lake to Old Forge via Long Lake, Blue Mountain Lake, and Raquette Lake.

A question absolutely essential to any contemplation of the constitution and wilderness and how both have fared in the courts is this: what did the addition of "trees and" signify? Delegate Dow noted, first, that it was intended to clarify the implied intent of the 1894 convention in its adoption of the word "timber," thus acknowledging that to some observers the word "timber" might exclude what would henceforth be more comprehensively defined as trees *and* timber. Further, the proposed commission would have the power to construct fire trails and to remove dead trees, "for purposes of reforestation and fire protection solely."[7] He argued that such was the intent of the 1894 convention.

The recommendation from the Conservation Committee that the State be permitted to clear fire trails in the Forest Preserve introduced the sort of ambiguity that has confused matters throughout nearly all of these documents. How wide would such trails be? What is the difference between a "fire trail" and a "fire road"? How much tree cutting would be permissible? In any case, the assertion that they would be laid out for fire protection only and that they were explicitly to be "trails" suggests a distinct limit to their purpose and their design. They were not to be constructed or maintained for recreation or anything other than planting trees or suppressing fires. They were not to be opened to the public. They were not to be roads for trucks, cars, or other motorized vehicles.

In the discussion that followed Delegate Dow's introduction of his committee's recommendations, we find lengthy rehearsals of other matters, important at the time: the precise function of the proposed commission, whether or not the State should be empowered to lease campsites, and how the State should settle the issue of clouded title on lands around Raquette Lake that it believed were state owned but on which families had occupied houses and camps for decades.[8]

After considerable attention to these and other matters, especially how the proposed Conservation Commission was to be appointed and overseen by the legislature, with Louis Marshall (father of Robert Marshall) leading most of the debate, the convention returned to the language of the proposed Article 7, Section 2. The new language, with "trees and," was read aloud to the delegates again, precisely as above.[9]

The chief opposition to the newly proposed Section 2 was offered by Delegate Angell, who we learn elsewhere was an attorney for the Santa Clara Lumber Company, one of the largest private landowners in the Park and a major player in the Adirondack pulp and paper business.[10]

Angell proposed that the entire Forest Preserve be divided into two classifications, first, the mountaintops and lake and river shores, and second, all the rest, which would be open to logging. Angell further proposed that highways be permitted in the Forest Preserve and that the State be explicitly granted the right to lease campsites. This was the essence of the minority report from the Conservation Committee, the main opposition to the Conservation Committee's majority recommendation. Angell argued at length against the addition of "trees and." "This is a provision which, if possible, makes more inaccessible the Adirondacks and the Catskills to the people of New York than they have ever been before." He argued that if this were approved, no one would be able to camp in the Adirondacks or "cut a tent pole, a tent stick, or anything in the Adirondacks for the purpose of the construction of a tent, or for any other use whatever."[11] This *reductio ad absurdum* fallacy has often been used by opponents of any rigorous interpretation of the forever-wild provision. But it is crucially important to note that when the convention rejected Angell's argument, as it overwhelmingly did, and adopted "trees and," it was explicitly tilting toward an understanding of constitutional intent that protected everything, not just those trees large enough to be worth something on commercial markets. Delegate Angell's amendment was voted down. His was the only objection expressed on the convention floor to the addition of "trees and."

Following Angell, we see a series of speakers offering their thoughts on Article 7. These included delegates supporting the current constitutional protections and a few opposed. They repeated all the arguments from 1894, adding nothing new to the debate but occasionally waxing poetic. Almost exclusively, they pitched their positions in general terms—the importance of watershed protection, the value of the Forest Preserve as a recreational destination. No one explicitly noted how "trees and" would change the level of protection.[12]

The entire text of the proposed constitution was read into the record, including the new Section 2 of Article 7, appearing precisely as above, with the addition of "trees and."[13] In voting for or against the entire new constitution, many delegates objected to one thing or another, mostly to provisions regarding apportionment for legislative seats, to unfair treatment of New York City, or to a putative slide toward "socialism," a word used by several delegates as they explained their vote. Not one mentioned "trees and," and the new constitution was approved by a majority of delegates and sent to voters, with the state-wide vote scheduled for November 2, 1915.[14]

A critically important document relating to and explaining the addition of "trees and" to the Forest Preserve provision demands our scrutiny. This was an article in the *New York Times* written by Louis Marshall and published on Sunday, October 24, 1915, nine days before the state-wide referendum on the proposed constitution.[15] Louis Marshall was a delegate at both the 1894 and 1915 conventions. He sat on the Conservation Committee at both conventions and was a staunch defender of the Forest Preserve. He was also a lawyer who argued complex cases before the US Supreme Court and the New York Court of Appeals.[16] He thus was especially sensitive to the importance of precise language and understood well the need for writers of constitutional provisions to strive to minimize ambiguity, vagueness, or imprecision.

Marshall noted that the new provision, including "trees and," was approved at the convention by a vote of 121 to 11. He further asserted that the chief goal of the Conservation Committee was the "conservation of the natural resources of the state." More important, Marshall explained concisely why the convention added "trees and." The reason for this addition was that the original language adopted at the 1894 convention and subsequently approved by voters might suggest "doubt . . . as to the comprehensiveness of the prohibition against the sale, removal, or destruction of timber." To remove this doubt, Marshall explained, the 1915 convention added "trees." He continued, "The new provisions emphasize and strengthen the efficacy of the existing prohibitions." Finally, he declared that the new language prohibited the removal of a "twig of a tree or a single drop of water" from the Forest Preserve.

Given the arguments advanced by the State in defense of tree cutting for snowmobile trails a century later in *Protect*, these remarks by Louis Marshall are critical to our understanding of the original intent of the 1894 convention. Marshall is arguing, forcefully, that that convention meant to protect far more than timber of market value, that it sought to preserve inviolate every twig, stem, and shoot. We can debate endlessly (and have) whether this intent was sound or would prohibit such intrusions as trails for snowmobiles, which did not even exist in either 1894 or 1915, and we can decide that modern circumstances demand moving beyond that original intent. But we must begin by figuring out what that intent was. This article by Louis Marshall is crucial to that enterprise.

It comes as a bit of an anticlimax to note that the constitution written in 1915 was rejected by the voters. In addition to what for our purposes was a crucial clarification of the provision protecting wilderness in the Forest Preserve, the document as a whole provided a model of progressive reform, an effort to bring professional and scientific expertise into the day-to-day operations of a huge—and growing—state government. No less an authority than the progressive historian Charles Beard observed that it introduced "a degree of responsible government hitherto unknown in American politics."[17] This was too much for voters, although many of the reforms were later implemented legislatively. But the gloss offered by the convention—and Louis Marshall, in particular—on the original understanding of Article 7 was invaluable.

19

MacDonald

The first important judicial examination of how the constitution protected the Forest Preserve occurred with *Association for the Protection of the Adirondacks v. MacDonald,* the final decision for which was handed down by the Court of Appeals in March 1930.[1] The story of this decision is a well-worn chestnut of Adirondack history. Once Lake Placid had won approval to host the 1932 Winter Olympics, it needed to find a place for a bobsled run. The Lake Placid planning committee thought the best site was on state land, in the Sentinel Range, and the legislature agreed, passing enabling legislation in 1929. The Association for the Protection of the Adirondacks, at that time the only environmental lobby paying much attention to the Adirondacks and, more particularly, to how the Forest Preserve was protected by the constitution, believed that construction of this facility, which required cutting many trees, would be a clear violation of Article 7 and sued the State to stop it. Litigation wound its fitful way through New York courts until reaching the Court of Appeals, which agreed with the Association and tried—with mixed results—to clarify matters. The 1930 decision remained the only significant intervention from the state judiciary until *Balsam Lake,* which was settled in 1993.

The Court of Appeals found that the completed bobsled run would "necessitate the removal of trees from about four and one-half acres of land, or a total number of trees, large and small, estimated at 2,500." The court acknowledged the vast extent of the Forest Preserve and admitted that "the taking of four acres out of this vast acreage for this international sports' meet seems a very slight inroad upon the preserve for a matter of such public interest and benefit to the people of the

122

state of New York and elsewhere." Further, the court noted, the State appeared to have the right to do whatever was needed to prevent fires and provide for "the erection and maintenance of proper facilities for the use of the public," provided that these "did not call for the removal of the timber to any material degree."

That phrase, "to any material degree," is a problem, offering the wiggle room and ambiguity that seems inevitable in such matters. I am guessing, without hard evidence, that the court had in mind here the state campgrounds that had appeared in the Adirondacks and Catskills after the first World War, the fire-lookout towers on summits, hiking trails, and the lean-tos built for backpackers and paddlers on state land.[2] All of these involved some measure of cutting, but the court was drawing a line, saying that this level of cutting was permitted, that the legality and reasonableness thereof was implied in the provision of 1894.

But what about the clearing of 2,500 trees from four and a half acres, for an activity that would, by all appearances, offer recreation and excitement to New York citizens and tourist-driven revenue to businesses in and near Lake Placid? What, the court asked rhetorically, could constitute an objection to such an incursion (similar to that needed for a campground like the one at Fish Creek Ponds, which the court did not specifically name but which I believe it considered) into the state's forests? "One objection, and one only—the constitution of the state, which prevents the cutting of the trees." Note: the court said "trees," not "timber." The court understood the word "timber," as used in Article 7, to include all trees, not just those of market size. After quoting Article 7, Section 7, the court pursued this point: the constitution insists that "the timber, that is, the *trees* [emphasis added], shall not be sold, removed or destroyed."

The court was well aware that constitutions involve language and that language demands interpretation, declaring (in a turn of phrase that crystalizes much of the message of this book), "Words are but symbols indicating ideas and are subject to contraction and expansion to meet the idea sought to be expressed." Looking at the history of the Forest Preserve and the conditions that led to the prohibition of 1894, the court observed that the delegates at the convention of that year believed it "necessary to prohibit any cutting or removal of the *trees and timber* [emphasis added] to a substantial extent." With "to a substantial extent," the court was trying valiantly to be reasonable but condemning future policies about cutting to the same uncertainty that led to this case in the first place—and to others, namely *Balsam Lake*, followed by *Protect*.

The court ran through the history of amendments permitting the construction of state highways and inferred that the fact that amendments and their complicated and time-consuming mechanics were required showed that the intent of Article 7 was manifest: "Trees could not be cut or the timber destroyed, even for the building of a road." But then, in a dodge that would get us back into the courts in the twenty-first century, the court acknowledged that it could not settle the matter definitively when it came to cutting for recreation: "What may be done in these forest lands to preserve them or to open them up for the use of the public, or what reasonable cutting or removal of timber may be necessitated in order to properly preserve the State Park [*sic*] we are not at this time called upon to determine."[3] It did not know exactly what level of cutting was "reasonable." But it was willing to declare that the amount of cutting predicted for the bobsled run was too much and was thus unconstitutional. Continuing with its insistence that the constitution protected trees in general and not just big trees, the court declared that the delegates of 1894, in order "to stop the willful destruction of trees upon the forest lands, and to preserve these in the wild state now existing, . . . adopted a measure forbidding the cutting down of these trees to any substantial extent for any purpose." And there the uncertainty creeps back in, with "to any substantial extent."

MacDonald accomplished much, but the job remained unfinished. The court declared that there was a constitutional threshold concerning cutting. Removing 2,500 trees from four and a half acres was beyond that threshold and illegal. The court did not say that cutting only 1,000 trees would be legal, or 2,000 trees, or 2,499 trees. It declined to define the word "tree." Was a young tree fully a tree in its first summer? Did it become a tree by surviving its first winter? Did it need to achieve a certain height or diameter to be considered a tree? Fifteen feet tall? Twenty? An inch diameter? Three inches? The New York Courts were not finished with the Forest Preserve. And the court's implicit concern about state campgrounds struck certain delegates to the next constitutional convention as needing attention.

20

The Constitution, 1938

The action concerning the Forest Preserve at the 1938 convention did not occur on the convention floor, in discussions involving all the delegates, but in committee, where interested parties not serving as delegates could offer testimony and where a serious effort to change the wording of Article 7 was proposed and defeated.[1] Notwithstanding vigorous debate at the Conservation Committee about the future of the Forest Preserve and how it should be protected by the constitution, the only change to come out of this convention was a different number for the forever-wild provision; Article 7, Section 7, became Article 14, Section 1. Wording and punctuation remained exactly the same.

Adirondack historians (including Alfred Donaldson, Frank Graham, and myself) have closely examined the conventions of 1894, 1915, and 1967, where there was active floor debate on the Forest Preserve. Curiously, the convention of 1938 has been largely overlooked. In 2016, blessed with a travel grant from the New York State Archives, I drove to Albany and spent three days plowing through the records of that convention: boxes of papers, correspondence, committee minutes, newspaper clippings, and the various notes and ephemera generated by a modern, bureaucratic state trying to improve its governance while also answering the whims of partisan politics.

Between 1915 and 1938, several amendments to the forever-wild provision, mainly for roads, had won the voters' approval.[2] Meanwhile—and more important—a series of decisions from state attorneys general had pushed against the stricture forbidding the destruction or removal of trees. These gave the State the go-ahead to develop recreational facilities on the Forest Preserve. State campgrounds, for example, required the removal of trees. Particularly after World War I, when the conservation bureaucracy began to see its chief role in the Adirondacks (and Catskills) as a facilitator of recreation, the Conservation Commission (renamed the Conservation Department in 1927) aggressively stretched the constitutional prohibition.[3] The fledgling conservation lobby, chiefly the Association for the Protection of the Adirondacks, grumbled, but no one was eager to take the State to court for its predilection for cutting down trees on the Forest Preserve.

This changed with *MacDonald*. But as has become increasingly clear ever since, the court's decision, a critical phrase of which prohibited "the removal of the timber to any material degree," left too much room for competing interpretations, and the State continued to push recreational developments that nibbled away at the constitutional restriction. As the 1938 convention approached, both defenders of the Forest Preserve and those hoping to see even more substantive state development for recreation expected the delegates to clarify the forever-wild provision of 1894.

The process began in November 1936 with a statewide referendum. Overshadowed by the presidential election that year, with little notice from either political party, with virtually no public education, and with the *New York Times*, among others, editorializing against a convention ("No Convention Needed Now," declared a headline in the *Times*), the vote generated little interest. Of all New Yorkers who voted in the Roosevelt-Landon presidential contest, fewer than half registered a preference on the convention question. The final tally approved it, but there was no enthusiasm and minimal awareness of the stakes.[4]

Primaries for delegates were held in September 1937. In the final vote for delegates in November with no other significant issue on the ballot, voter turnout was low. But once it was clear that a convention would in fact sit, the political implications, which inevitably included a debate on the always-contentious subject of apportionment of legislative districts, were obvious, and both parties geared up for the convention. The Democratic party, with its strength largely in New York

City, would control the convention and would push a list of reforms, but for both major parties the Forest Preserve was way down the list of pressing issues. Depression-era poverty and the agenda of the New Deal were the dominant topics of the day, leaving any changes to the Forest Preserve provision to the attention of a handful of interested parties, who expressed themselves almost exclusively at the committee level.[5]

In correspondence to the Conservation Committee, at the only public meeting it held, on June 23, 1938, and in the committee's report to the convention as a whole, three possibilities dominated all discussion of the Forest Preserve: (1) keep Article 7 in exactly the same language; (2) implement a multiple-use policy (as in the National Forests), with commercial logging and all sorts of recreation; or (3) maintain the prohibition on logging but rewrite Article 7 with an explicit clarification allowing the Conservation Department to develop state lands for recreation.[6] When the Conservation Committee held its only public hearing, where anyone with an interest in the future of the Forest Preserve could speak and submit remarks to be included in the committee's official record, these were the three positions on which virtually all participants focused.

Hundreds of letters from private citizens, high-school biology classes, garden clubs, the Appalachian Mountain Club, and many others made the case for no change in Article 7. The New York Chamber of Commerce, for example, passed an official resolution supporting retention of the exact language written in 1894. The substance of these letters, as well as the nature of the organizations submitting them, shows that in a matter of a few decades, especially since the convention of 1915, the popular understanding of the value of the Forest Preserve had come to focus almost exclusively on recreation, on the spiritual importance to the state of rigorously protected wilderness.

Among those few advocating opening up the Forest Preserve to commercial logging were professional foresters and representatives from forestry schools such as those at Yale and Syracuse. Correspondence from Clarence Fisher, of Fisher Forestry and Realty Company, was characteristic: the current policy, wrote Fisher, "causes industrial stagnation and by taking all the timber of the State entirely out of the market shuts down wood-using industries and deprives local residents of the opportunity for work." This was a familiar argument, routinely expressed by logging companies ever since the Forest Preserve was granted constitutional protection in 1894. It was also incorrect: in addition to the substantial logging taking place on private land, the State had already begun acquiring what would become the system of state forests, where logging is permitted, scattered throughout New York, administratively distinct from the Forest Preserve and not subject to the protections of the constitution.

Speaking at the public hearing in favor of no change were Lithgow Osborne, the state conservation commissioner; John Apperson of Schenectady, a tireless defender of the Forest Preserve; Bob Marshall, representing the Wilderness Society; and many others, all urging the committee to maintain the forever-wild provision in precisely the same language. The committee also heard representatives from the logging industry, but it's clear from the record that there was no measurable interest in their arguments. The committee overwhelmingly rejected commercial logging on the Forest Preserve and spent little time even discussing it. Along with the similar rejection of pleas from commercial logging interests in 1915, this quick brush-off of the idea of opening the Forest Preserve to logging shows clearly how the State had pivoted from the widely held position of 1894, where it was assumed that the prohibition could be reexamined in a few decades. The watershed argument had faded, replaced in the popular consciousness by the idea of the Forest Preserve as a wilderness retreat for all. The only real question was how much latitude the State had to facilitate recreation.

The serious discussion that day—a discussion which has dominated our understanding of the Forest Preserve and its inscription in the constitution to this day—was about clarification: was the State violating the constitution when it cut down trees for recreational facilities, and, if so, should the constitution be changed to indicate explicitly what the State could and could not do? In other words, spurred to action by the decision in *MacDonald*, the convention was looking at precisely the issue that led, many decades later, to *Balsam Lake* and *Protect*. Interestingly, Lithgow Osborne, whose Conservation Department was even then enthusiastically doing the problematic cutting—for campgrounds, fire towers, lean-tos, and the like—led the fight against clarification. Such a change, which he said was designed to "give a constitutional validity" to present policies of the department, was "unnecessary." Throughout Osborne's lengthy statement of opposition to any clarifying change was this fundamental position: with no plans to promote anything as intrusive as a bobsled run, now identified as unconstitutional, and with a succession of opinions from attorneys general giving the Conservation Department carte blanche (more or less) on other sorts of recreational developments, the State believed that any fiddling with the constitution would only muddy the waters. In other words, suggested Osborne, *what we're doing is legal because we say it is, so don't rock the boat*. To this day, this attitude has prevailed at the upper levels of the state's conservation bureaucracy.

Osborne acknowledged that what he was saying was "debatable." But, he speculated, "I do not believe any court would interpret Article 7, Section 7 as a dead hand constraining our people from any recreational use of the preserve which would leave unaffected its essential character." His definition of reasonable recreational pursuits included anything "which primitive people did of necessity in a wilderness and which we do now for sport"—a definition, we should carefully note, that would preclude snowmobiling or even bicycles anywhere in the Forest Preserve. Such activities, i.e., those "which primitive people did of necessity in a wilderness," are "to be encouraged in the forest preserve even when trees are to be destroyed in making them possible, provided of course that, to quote the Court of Appeals, 'they do not call for the removal of the timber any material degree.'" To change the status quo, Osborne argued, would be "not only unnecessary but may even act as a bar towards other equally permissible but unmentioned recreation activities." He was certain that the State could be trusted to make reasonable decisions when it came to cutting down trees in the Forest Preserve, and he preferred that constitutional clarification not be sought—at least not at the 1938 convention. This was exactly the assumption that led the State to undertake the construction of snow-mobile trails litigated in *Protect*.

The argument for making precisely the sort of change that Osborne so strenuously argued against was eloquently advanced by Theodore Cross of Utica, who pointed out the creeping incursions of the state on Article 7. Cross's message to the Conservation Committee neatly distills the debate and anticipates *Protect*: "A private citizen cannot go upon the Forest Preserve lands and cut a tree—perfectly proper construction of the Constitutional requirement; on the other hand, why should . . . the New York State Conservation [Department] be permitted to go upon the Forest Preserve to cut trees, to construct highways and to prepare public campsites?" Cross was not opposed to such cutting by the State. To the contrary, he went on to insist that Article 7 was "an out-moded provision, absolutely unusable in its present form. The provision of the Constitution should be so changed as to permit the reasonable adaptation of the Forest Preserve to the present requirements of the People." In other words, without such explicit clarification the current Conservation Department policy was unconstitutional. Cross was thus no friend of wilderness, but he clearly understood the constitutional stakes. As we shall see, the decision of the Court of Appeals in *Protect* echoed what Cross observed in 1938: if the State wants to cut down trees for roads, the constitution must be amended.

Others jumped in, but nothing new was added to the irreconcilable positions of Osborne and Cross. To Osborne the state policy was "debatable" but eminently reasonable, while to Cross and others it was illegal.

So what did the convention do? Nothing. Osborne was the conservation commissioner, the voice of authority, while Cross and others were merely private citizens unhappy with the State's willingness to stretch the constitution. The delegates went along with Osborne. And for several decades, environmentalists occasionally grumbled, as they did, for example, when the State built horse trails and barns in Cold River Country in the western High Peaks in the late 1960s.

The rest of the story is quickly told. The Conservation Committee reported the original 1894 provision for the Forest Preserve to the convention with "not even a comma . . . changed." All the convention did was provide new numbers. Where previous constitutions had been presented to voters for one up-or-down vote, the constitution of 1938 was submitted in nine packages, requiring nine votes. The first of these, which included the Forest Preserve article, was approved, though not by a large margin, while others (reapportionment, for example), more controversial, were defeated. All the amendments received far fewer votes than did major elective offices. In December the *New York Times* declared the whole process a disaster and a waste of money. Voters were confused by the nine separate lines on the ballot, and many cast "blank or void ballots" on one or more of them, to the extent that these "blank and void" ballots outnumbered yeses and nos.[7]

21

The Constitution, 1967

At the constitutional convention of 1967 a debate very much along the lines of that in 1938 recurred.[1] Were large campgrounds constitutional? Although the delegates routinely and confusingly referred to what they called "campsites," the issue was the large, state-developed sites where families could set up a tent near their car or stay in a motorized recreational vehicle. The debates did not address primitive sites deep in the woods where the state offered lean-tos, a fireplace, and privies; no one questioned the constitutionality of these. The debate concerned campgrounds like the one at Fish Creek Ponds. If they were not constitutional, or even if their constitutionality was ambiguous, should the language defining and protecting the Forest Preserve be changed? No one, at least no one on record, wanted to prevent the State from clearing or continuing to maintain a reasonably sized spot on a lakeside for drive-up camping. The state-run campgrounds—where urban or suburban families could spend a week or two beside a lake with clean water, near a town with restaurants and grocery stores, using a safe brick fireplace and community showers—were enormously popular. The problem was how to eliminate the embarrassment of maintaining facilities that might be technically forbidden by the state constitution.

Constitutional amendments had been deemed necessary for state-run ski slopes—Whiteface and Gore in the Adirondacks, Belleayre in the Catskills—and by the 1960s, the State was maintaining not only campgrounds, but wide trails to accommodate horses.[2] Just as in 1938, questions about constitutionality persisted, particularly since nearly all observers were certain that without amendments to the constitution, the ski slopes could not have been developed. What, they wanted to know, made the campgrounds different?[3]

Before the matter of campgrounds was addressed, however, the convention dealt, as it had in both 1915 and 1938, with a proposal to radically modify forever wild and adopt the sort of multiple-use policies governing management of the National Forests, in other words an abandonment of the idea of protecting wilderness. This was introduced by Delegate Francis Bergan of Albany, a Democrat, who had been a delegate to the 1938 convention and was a justice on the New York Court of Appeals. Bergan's aim was to expand significantly the sorts of recreation that the State could legally promote. Bergan declared that he wanted to maintain the words "forever wild," and his description of what he foresaw under his proposal included "sound conservation management." But what he intended was development of recreational facilities far beyond what was currently permitted, including public roads into the High Peaks, ski slopes on Algonquin, and state-managed hotels. He also wanted to see forest management to cull mature trees and thus promote the deer population by opening the canopy. Whether he supported the sort of logging contracts and significant commercial harvesting that the National Forests allowed is unclear.[4]

Opposition to Bergan's proposal was immediate and forceful. Erastus Corning, the mayor of Albany, issued a stirring defense of wilderness and suggested that Bergan obviously did not even know what "forever wild" meant:

> When he says that that is what he means by Forever Wild, it isn't what I mean by Forever Wild and I don't think it is what the people of the State of New York mean by Forever Wild. I don't think that they want the Adirondacks managed meticulously as a forest. I don't think they want the best game management. I don't think they want it full of recreation. I do think that the people of the state of New York want to see this greatest of all state parks maintained as it has been for over 70 years as Forever Wild forest lands, and I hope that this amendment does not prevail.[5]

Corning's objection to management designed to promote deer and open up the Forest Preserve to recreation other than backcountry activities like hiking or canoeing is especially telling; it was a defense of wilderness as such, an echo of the arguments advanced three decades earlier by Robert Marshall.

Also objecting to Bergan's plan was State Senator Watson Pomeroy, a Republican, who represented a district in the Hudson Valley and had spent the last several years as a member and then the chair of the Joint Legislative Committee on Natural Resources; he was probably more familiar with the Adirondacks and the Forest Preserve than any delegate at the convention.[6] Pomeroy zeroed in on Bergan's effort to turn the Forest Preserve from wilderness to a managed recreational playground, reminding delegates that New York already had an extensive Park System with amenities and facilities like those advocated by Bergan and that the Forest Preserve was different precisely because it offered wilderness:

> [The] forest preserve does furnish the only wilderness type area east of the Mississippi River, with the exception of the relatively small Baxter State Park in Maine; the only place where people can go for hunting, fishing, hiking, and camping, in natural surroundings in a great deal of this area. Not all, of course. No other state has this, no other state can have this because New York State, in the wisdom of the people in 1894 decided that we were going to keep this last vestige of remaining wilderness. . . . That is what we are talking about when we talk about the forest preserve: something that is unique, something that is used . . . by millions of people and something that is unique because in all the eastern United States, New York State is the only one with this valuable asset, and I trust that this amendment will be defeated.[7]

Speaking in favor of abandoning the tight strictures of forever wild was, among others, delegate Neal McCurn of Syracuse, who spelled out in unmistakable detail the class-based element in the debate over whether or not New York should continue to protect the Forest Preserve as wilderness. McCurn insisted that advocates of keeping Article

14 just as it was did not live and work in the Park, while all those in favor of the looser management policy proposed by Bergan lived in or close to the blue line. McCurn rather sneeringly pointed out that the Wilderness Society, on record as supporting Article 14, was based in Washington, DC. The notion that Bergan's proposal was more sympathetic to the wishes of the year-round residents, who ostensibly wanted more tourists and less wilderness, was likewise advanced by Delegate Claude Clark of Franklin County. Clark also challenged the constitutionality of the campgrounds, asking, "And I wonder, are these public campsites [sic] in the Adirondacks actually legal?"[8]

To McCurn, as to so many others who cannot see the value of letting trees mature and die and return their substance to the soil, the Forest Preserve was a colossal mistake: "You can't preserve a forest by allowing it to go on and grow one tree on top of another, or a primeval forest, as they call it. The only way to preserve it is to cut it and have second and third growth timber in there." The alternative was "an inaccessible area which is of no benefit to anybody."[9] In the statements of Erastus Corning and Neal McCurn we have a distillation of the classic and endless argument about wilderness. To Corning it was invaluable, a treasure to be guarded in perpetuity by New York and its constitution. To McCurn, it made no sense at all. Interestingly, Corning, McCurn, and Bergan were Democrats, while Pomeroy and Clark were Republicans. Henrik Dullea has shown that while nearly every issue debated at the 1967 convention was settled along more or less rigid party lines, this was not the case with the Forest Preserve.[10] The convention overwhelmingly voted down Bergan's proposal, 152 to 18.[11]

But Article 14 was not off the table. Defending the Bergan proposal, McCurn and others, aware, no doubt, of how this subject had vexed the 1938 convention, had touched on the issue of the campgrounds and their constitutionality. McCurn declared, "We presently have camping sites in the wilderness areas and these are available to the citizens of the state. I think it goes without saying that these camp sites that are there now are unconstitutional because they are in violation of the law."[12]

Immediately after the vote on the Bergan proposal, Charles Froessel, Democrat from Brooklyn, recently retired from the New York Court of Appeals, where he had served from 1949 to 1963, returned the floor discussion to the campgrounds and their questionable constitutionality. Judge Froessel proposed to add words to Article 14 to explicitly permit "the construction, maintenance and operation of recreational campsites bordering on or in the vicinity of state or county highways with necessary access, water supply and sanitation facilities, all of which shall be in keeping with the surrounding areas." Explaining his aim, he argued that the current wording of Article 14 did not permit recreational facilities and that the Conservation Department had

> stretched the constitution and they have erected not only within the Adirondack Park system, but within the blue line, at least 40 campsites during the last three or four decades. There is no basis in law for their having done that. . . . Now, I am not criticizing the commissioners in the past for doing this, because they have only been trying to meet the demands of the people of New York to enjoy their own forest preserve. But at least there should be a legal foundation for it.

He quoted Pieter Fosberg, editor of the Conservation Department's magazine, the *New York State Conservationist*: "Almost everyone believes campsites are necessary and desirable, but which are certainly of doubtful constitutionality." Froessel continued,

> My amendment would serve to legalize or give a foundation to those campsites that have already been erected by the conservation commissioners in the past and would also give them an opportunity of creating further campsites to accommodate the great number of people who are turned away or have to wait at the peak periods during the summer time period.[13]

Then, in a dramatic moment that has become an often-told anecdote in the history of the Forest Preserve, Delegate Dollie Robinson, a Black Democrat from Brooklyn and member of the committee that had recommended keeping Article 14 in its original language, rose to object to Froessel's proposal. Robinson, who began her remarks by acknowledging that she had never set foot in the Forest Preserve, had not, in fact, even seen the Forest Preserve except perhaps for a glimpse of a Catskill summit from the Thruway, argued that the convention should entertain nothing that diminished the strict protections already in place. She hoped that people like her—urban, often poor and unable to get from their homes to the Adirondacks—would someday have the opportunity and that their experience would be just as moving as it had been for those more privileged. When she finished, the convention chair had to remind the delegates that applause after speakers for or against amendments was prohibited.[14] It remains to be seen whether the Adirondack wilderness can in fact be made both available and welcoming to New York's non-White residents, whose tax dollars continue to pay for managing the Forest Preserve and to support Adirondack schools and local government.

Others objected to the Froessel amendment, including Watson Pomeroy. Raymond Rice, Republican from Nassau County, offered an odd and convoluted assessment of the constitutionality issue:

> We don't have to legalize these campsites which have existed over the past 40 years and on this basis we would again take issue with Judge Froessel because if they were not legal they have existed since 1920 and it would certainly mean that many, many of our public officials have been operating under a false flag and that the legislators, our Governor, and all those who have been at all concerned with the operation over all these years have been fooling the public.

Rice thus more or less adopted the position urged by Lithgow Osborne in 1938: "This amendment is not needed, moreover it's not wanted."[15] Watson Pomeroy's position was similar: the campgrounds were a fact, the Conservation Department saw no need for an amendment explicitly legalizing them, and the people of New York were enjoying them.

After much give and take behind the scenes and thus not on the official record but described by Frank Graham, who was able to interview key actors, Judge Froessel's proposal was incorporated into a proposed new constitution. For the second time (the first was 1938), the Forest-Preserve provision was moved to a new article, this time to Article 8, where Section 1 repeated precisely the language of 1894, with not a word or a comma changed. But a new Section 2 listed exceptions, areas where the principle of forever wild and the prohibition on removing any timber did not apply. These exceptions mostly involved activities for which amendments had already been approved, e.g., the provision for dams needed for flood control, passed in 1913, or provisions for state highways. The critical addition for our purposes involved the campgrounds: excepted from Section 1 was "the use of such lands for public campsites of the kind presently constructed and maintained, and in areas similar to those in which they are presently located."

Environmental groups were not enthusiastic about this, presumably because any foot in the door for a diminution of the authority of the constitution could be just the beginning of an end to constitutional protection of wilderness. But, according to Graham, no one had the stomach for disputing Froessel, a man universally admired and respected. He was a legal scholar who believed in taking constitutions seriously, and in this case he was right: in nearly any reading of the constitution, including the discussions at the conventions of 1894, 1915, and 1938, the campgrounds were illegal. To Judge Froessel, it was more important to make the constitution consistent with reality than to protect wilderness at any cost. Froessel and the convention delegates, unlike their predecessors in 1938, were implicitly acknowledging that the campgrounds were illegal and had been constructed by the Conservation Department in violation of the constitution.

The constitution of 1967 was rejected by New York voters in November, mainly because of widespread opposition to a provision permitting state funds to be funneled to parochial schools. The 1938 revisions had been offered to voters in several different clusters: voters could approve the changes they liked and reject the ones they didn't like. Consequently, not all the revisions of 1938 were approved. The 1967 convention resolved, in a decision that has been widely criticized since then, to offer voters a substantial new constitution for one up-or-down vote. The provision concerning state aid to parochial schools poisoned the debate, and the entire constitution, including Judge Froessel's amendment, about which most voters probably cared not a whit, went down.[16] So the status of the campgrounds remained ambiguous. As far as I'm concerned, they are illegal, but I understand that taking the State to court would be political suicide for any organization opting to contest their constitutionality. I also would guess that something akin to the doctrine of adverse possession might apply: the principle of state-maintained campgrounds on Forest Preserve land is over a century old, and the window for reasonably challenging them closed long ago.

But the decision to try retroactively to legitimize the campgrounds sheds considerable light on how the delegates of 1967 understood the work of the delegates of 1894, the decision in *MacDonald*, and the inconclusive debate in 1938. The delegates of 1967 believed the campgrounds violated the constitution, as it was written in 1894 and interpreted by the Court of Appeals in 1930. Unlike the delegates in 1938, who may indeed have believed the same but declined to act, the 1967 delegates tried to insert their understanding into a proposed new constitution. They declared that if the State wanted to cut down trees for anything more substantive than a hiking trail, it must secure an amendment. Their understanding of the forever-wild provision, of course, is not determinative. It does not possess the authority of the constitution itself or of a ruling in the courts, but it is an important part of the historical record as we approach the decision in *Protect*.

22

Balsam Lake

Between the 1967 convention and the final decision in *Protect*, much happened in the developing story of wilderness in the Forest Preserve. In 1971 the New York legislature created the Adirondack Park Agency and charged the new agency to draw up a Master Plan for the management of state land in the Adirondacks. The primacy of wilderness and its protection in the constitution is vigorously asserted in the State Land Master Plan (SLMP), which was approved by Governor Nelson Rockefeller in 1972, but it added little to continuing efforts to figure out just what the constitution means.[1] Notwithstanding the lack of clarity, however, beginning in 1972, the idea of protecting wilderness, *as such*, became official state policy. Wilderness had been on the table since 1894 and was discussed at every constitutional convention. But with the SLMP, the notion of wilderness as a locus for recreation and spiritual revival enjoyed the imprimatur of explicit official policy.

The next chapter in this story surfaced in litigation, this time in the Catskills. As with *MacDonald*, the dispute arose when the State of New York proposed to cut trees on the Forest Preserve, more trees than a litigant believed were constitutionally permitted, to develop recreation—the same thing that happened later in *Protect*. Believing that the state bureaucracy was violating the constitution, a New York organization sued. And each time, the courts offered an interpretation of the constitution that refined, although imperfectly, our understanding of how the constitution protects wilderness and governs what the State can and cannot do with what the constitution declares "shall be forever kept as wild forest lands."

Each constitutional convention adds to the record, each decision in the courts does the same, and some (but not all) of the imprecision inherent in Article 14 diminishes. Debates at constitutional conventions and decisions from the courts are both the holy scriptures and the commentary of the sages with respect to the Forest Preserve. Like medieval monks we scrutinize these sacred texts and hope we are inching closer to revealed truth, always alert to the ambiguities inherent in language, the inevitable inconsistencies of verbal communication, and the intrusion of personal bias.

In 1985, well after the SLMP was adopted for the Adirondacks, the State began constructing a similar document for the Catskill Forest Preserve. The plan's final language was approved in 1989.[2] For the preliminary and final drafts of these plans, it should be emphasized, the constitution provided the bedrock, while the plans provided definitions and specific management guidelines, explicitly acknowledging that nothing in a state management document can go beyond what the constitution allows. Because the constitution continued to demand that "nor shall the timber thereon be sold, removed or destroyed," nothing in the plans could change this fundamental provision. As always, the fulcrum of the new conflict involved interpreting what the constitution meant when it came to cutting trees: How many? Of what size? After 1930 there was one key document offering guidance on this in addition to Article 7 (renumbered in 1938 as Article 14): the decision of the Court of Appeals in *MacDonald* limiting how many trees the State could cut down in its effort to promote recreation.

In one section of the Catskill State Land Master Plan, the State proposed to clear certain acreage for a parking lot to accommodate hikers, anglers, and other persons seeking to recreate on state land and to reroute a trail for cross-country skiing. The Balsam Lake Anglers Club, which owned land adjacent to this unit of the Catskill Forest Preserve, believed the number of trees to be removed for the lot and trail went beyond what the constitution allowed and beyond the "any material degree" standard of *MacDonald*. They sued, and the litigation process quickly centered on what the constitution meant by "timber" and on how much of it could be legally cut by the State in its efforts to provide recreational opportunities for the public. Did little trees, less than an inch or two in diameter, count as "timber"? If so, how many little trees could legally be removed? This case and its resolution are thus clearly essential to our understanding—and the courts'—of the eventual decision in *Protect*, which came down some three decades later.

The case began in the Supreme Court of Ulster County, which issued its decision in 1991. The Balsam Lake Anglers Club argued that cutting "as many as 2000 'trees,' most of which are less than three inches diameter at breast height, constitutes the removal or destruction of timber" and thus violated Article 14.[3] The court examined the precedent in *MacDonald* and found that it had "rejected the absolutist argument that not even a single tree or even fallen timber or deadwood could be removed and stated that the constitutional provision must be interpreted reasonably." Such an absolutist approach, the court argued, would prevent reasonable use of state land by the public. "It is thus clear that the Court of Appeals [in *MacDonald*] determined that insubstantial and immaterial cutting of timber sized trees was constitutionally authorized in order to facilitate public use of the forest preserve so long as such use is consistent with wild forest lands."

The familiar ambiguity thus surfaces with "consistent with wild forest lands." But the Ulster County Supreme Court believed that this was the only viable way forward. Examining the record of the 1894 convention, it asserted, "There is no indication of any intent to maintain the forest in an 'absolutely' wild state with no organized human alteration or intervention at all." The Forest Preserve could constitutionally and logically be "kept as wild forest lands" and also have trails for hiking and skiing. The Supreme Court further ruled that the DEC had failed to follow strictly the New York State Environmental Quality Review Act (generally referred to as SEQR) and ordered the DEC to take steps to accommodate this omission. But in its finding relative to the constitution, it found the case of the Balsam Lake Anglers Club to be without merit: the level of tree cutting proposed by the State was legal.

Both the club and the DEC appealed, the club appealing the ruling on constitutionality and the DEC appealing the ruling regarding

SEQR. The case moved to the Appellate Division, Third Department, which issued what was the final decision in *Balsam Lake* in 1993.[4] This court upheld the lower court's interpretation of the constitution, beginning its assessment by acknowledging that *MacDonald* required a "reasonable interpretation" of the constitutional provision and that, consequently, the constitution did not prohibit all cutting. It also reversed the lower court's finding with respect to SEQR. The club thus lost on every part of its argument. It declined to appeal further, and the ruling of the Appellate Division became, along with *MacDonald*, a critical element in the constitutional saga.

The key element of the decision handed down by the Appellate Division found that the amount of cutting already completed or proposed by the State for parking lots and cross-country ski trails "is not constitutionally prohibited." In other words, where *MacDonald* had said *this much is too much*, the decision in *Balsam Lake* said *this much is acceptable*. The "this much" in *MacDonald* was 2,500 trees, "large and small," on four and a half acres, or about 555 trees per acre. Anything over that number is unconstitutional. The acceptable level in *Balsam Lake* was 350 trees on 2.3 miles of a ski trail.[5] Any cutting at that level or below it is constitutional. That means we still have uncertainty. How do we compare one measurement that uses acres and another that uses linear miles? Where is the crucial threshold between those two numbers? Equally important, moreover, after *Balsam Lake*, we still did not have an agreed-on understanding of what "timber," one of the key words in Article 14, means. To a cab driver in Brooklyn, this debate must seem endlessly pedantic and trivial. But to those of us immersed in the arcana of constitutions and wilderness, it is profoundly significant.

23

Wilderness and Constitutions

My job at BGSU was a good fit professionally, but I never cared for living in Northwest Ohio. Having grown up in West Virginia and with my heart in the Adirondacks, I found the perfectly flat expanses of corn and soybean fields in Wood County, Ohio, to be boring and depressing. Beginning on the gray, wintry day in 1980 when I first saw Bowling Green, I longed to be back in the East, in or close to the Adirondacks. A few years after my 2007 retirement from full-time teaching, my partner, whose own retirement was approaching, agreed that it was time for us to move to New York. The story of how we landed in Ithaca is another one of good luck. After a frantic few months of searching for a house across the eastern half of New York, I found myself settled in Ithaca, much closer to Long Lake and able finally to dive into the world of Adirondack environmental activism.

I had been minimally engaged while I worked at the Museum. It was the early 1970s, and after decades of not much happening in the Adirondacks, there was a frenzy of activism around the establishment of the Adirondack Park Agency (1971) and then around the Agency's two main contributions to open-space management in the United States: a plan for that part of the Forest Preserve in the Adirondack Park (1972) and one for private land in the Park (1973). Bill Verner was deep in the several environmental alliances promoting the Park Agency, and I followed him from meeting to meeting, sitting in the corner while Bill and others engaged in the nuts and bolts of nudging controversial and monumental legislation through the Assembly and Senate and to the Governor's desk. It was understood that Nelson Rockefeller was behind the entre initiative, but resistance was coalescing in the legislature. I listened with rapt fascination to Bill as he strategized with the likes of Dick Beamish, Liz Thorndike, Peter Paine, Marge Lamy, Courtney Jones, and many others.

My own contribution during those busy years included speaking up a couple of times at public meetings when the Adirondack Park Agency and its plans were on the table. At a hearing in Old Forge on the State Land Master Plan in 1972, I argued as well as I could for rigorous protection of wilderness, more than was currently on offer in the proposed SLMP. At a hearing on the Private Land Plan in Indian Lake in 1973—the same one where Long Lake businessman John Hosley assailed the Private Land Plan while wearing a faux Indian headdress and holding a toy spear and where Assemblyman Glenn Harris dramatically arrived by helicopter and announced his last-minute, and ultimately unsuccessful, efforts to delay the plan's implementation—I spoke in favor of it and was roundly booed.[1] I learned a few days later that at least one of my neighbors in Long Lake thought that since I did not at the time own property in the Park, I had no right to speak that night.

I moved out of the Adirondacks shortly thereafter, and my in-person involvement in contemporary politics ended—for the time being. But I was paying close attention to every Adirondack political moment while living in Ohio and then Michigan. Once back in New York, I had a fortuitous conversation with Michael Wilson that led to my joining the Board of Protect the Adirondacks!, a financially struggling but militant defender of Article 14 and its protection of wilderness in the Forest Preserve.

This, in turn, led to my involvement in what became a historic judicial proceeding, *Protect the Adirondacks! Inc. v. New York State Department of Environmental Conservation and Adirondack Park Agency*, Protect's suit against the state and its plans to construct certain snowmobile trails in the Forest Preserve.[2] When I joined the team arguing the case for Protect the Adirondacks, I did so for two reasons. One was a function of my personal disposition to wilderness and to protecting it. I hoped to contribute to making the case that Article 14 was at least in part designed to protect the wilderness I treasured and that its intent should be honored and extended. Another and rather different reason was my conviction that constitutions matter. Personally, I don't care for snowmobile trails cutting through the forest, and I especially don't like the noise. But resisting them in this litigation was only partly a function of my aesthetic preferences; if I thought the constitution had no position on these trails, I would probably have declined to participate.

Beginning with *MacDonald*, and running through *Balsam Lake* and *Protect*, the critical debate, as played out in the courts, focuses intensely on interpreting two sentences in the state constitution, to the point that the whole exercise can seem pointless to someone not deeply immersed in its history. But it is important because constitutions matter. In our American system constitutions stipulate how our elected and appointed officials interact with us; they organize what we do, how we do it, what our obligations to others are, how we declare what is significant and what is irrelevant, in short, how we organize all of our obligations to each other and to the state and the state's obligations to us. Our federal and our state constitutions are foundational to our version of civilized life. In New York, the constitution includes a fundamental provision determining how we, including our elected and appointed representatives, interact with a significant part of the natural world.

Composed of words, constitutions cannot avoid ambiguity and even circularity; they cannot be absolutely precise. But over the centuries, the traditions of English and American law have settled on a system that has worked reasonably well. As our history shows, it is a flawed system, unavoidably producing and reproducing anachronisms, uncertainty, and partisan pronouncements. It does not lead to perfect justice. But I can't imagine anything better. Constitutions and the elaborate mechanism by which we write them, approve them, amend them, and judicially interpret them are all we have. Dwelling on "original intent" often appears like a Jesuitical mind game, an exercise in nitpicking and pedantry, but, again, it is all we have. Establishing such a thing as original intent, of course, does not mean that once we think we have it, the process stops. Societies evolve, and constitutions can and should be amended. And courts should render their interpretations with changing circumstances in mind. But figuring out intent is a necessary first step.

All of this is relevant to why it seems so important to me to look closely at how our constitutions and their judicial interpreters have dealt with the protection of the Forest Preserve, how these have refined what we have come to think our constitution is saying. I understand that cutting down one small beech sapling of an inch or two diameter at breast height does not destroy the forest around it or even remotely eliminate the possibility that that forest can be reasonably understood to be a wilderness. But I also believe that when our constitution clearly states that cutting down trees is a constitutional issue, we do not cut them down casually. To do so would trivialize the document. If we make decisions about trees based on expedience, simple logic, or economic or even spiritual priorities without consulting the constitution, we are failing a test of responsible citizenship.

I am not arguing for a crude originalism, a locked-in effort to force all policies, laws, management goals, and the like into a procrustean obsession with words as deployed at some time in the past. Like wilderness itself, the notion of "original intent is a theoretical construct, not a fact in the world."[3] As the majority decision in *MacDonald* put it succinctly, "The words of the Constitution, like those of any other law, must receive a reasonable interpretation, considering the purpose and the object in view." But "reasonable interpretation" does not mean discarding the words in a constitutional provision in the heat of contingency or expedience. Those words are the essential starting point for assessing how a constitution forbids or allows one or another human activity. And, most important, when the state, as it did in *Protect*, insists that a sentence in a constitution means something specific, that its meaning is one thing and nothing else, and that that unchallengeable interpretation is fundamental to the legality of a perhaps controversial action, then we must go to that sentence and look at it carefully. If a society can reach a consensus that the meaning of a sentence can in fact be established and if, further, that precise intent can be agreed on as no longer in the best interest of a society, then that sentence can and should be amended. Such agreement is never unanimous, of course, but we have well-established—and justifiably cumbersome—mechanisms for amending our constitutions. As we shall see, that fundamental fact was critical to the final decision in *Protect*.

When the contesting attorneys in *MacDonald*, *Balsam Lake*, and *Protect* scrutinized the number and size of trees cut and when the justices considered their arguments and issued their opinions, they were all doing the best they could to honor a system that has served us as well as any system could. The constitution protects the forests and their timber, and the courts are there to tell us, to the extent they can, what that means. Cutting one little beech has been understood not to be a constitutional matter. Cutting every little beech, the courts tell us, is a constitutional matter. Finding the liminal point between those extremes is difficult. The process of interpreting the constitution looks for the tipping point between what the constitution permits and what it prohibits. In other words, the courts see the numbers and sizes of trees as the only way to resolve the debate. That is the issue lurking in plain sight in all three of these judicial decisions.

Like the mechanics of constitutional amendment, the history of motorized recreation in the Forest Preserve and discussions of its putative legality or illegality are also complicated. Indeed, wherever wilderness is protected on publicly owned land in the United States, disputes about whether or not motorized recreation can be permitted inevitably arise and are almost always bitterly contested. In the Adirondacks they are particularly fraught because they usually pit green groups, largely funded by supporters living outside the Park, against local politicians and business interests, who see snowmobile operators as important contributors to an always-sluggish winter economy.

In New York, moreover, such discussions are further complicated by the unique fact that much of our public land, the substantial acreage included in the Forest Preserve of the Adirondacks and Catskills, is explicitly identified as vital to the welfare of the state and guarded by the New York State constitution. So when we take sides about, say, snowmobile trails, it is more than a statutory or even an aesthetic issue. It is a constitutional issue, and that inevitably leads to court.

In our country, when we lapse into our frequently deep divides on constitutional issues, we focus intensely on meaning. It's a different sort of search for meaning than that of a bushwhacker on the slopes of Donaldson or of a researcher in the archives, but in my case at least, these meanings and efforts to understand them are inextricably connected. My explorations in the Sewards, my reading of Robert Marshall and William Cronon, my research in the records of New York's constitutional conventions—these and so much else all came together in *Protect*.

Because my research and publications appeared to qualify me as knowledgeable on the history of the Forest Preserve and because my involvement in the suit would be free, I was invited by Protect attorneys John Caffry and Claudia Braymer, who were working *pro bono*, to join the team as an expert witness. This meant helping them with research, testifying in the first round of the litigation at the New York Supreme Court in Albany in 2017, and trying to answer questions from Caffry and Braymer throughout the lengthy process as they wrote their briefs and responded to the state's briefs and *Amicus* interventions during the months and years that the litigation took to wind its ponderous way from its original filing to the New York Court of Appeals.

24

Protect the Adirondacks! Inc. v. New York State Department of Environmental Conservation and Adirondack Park Agency

Efforts to construct a gasoline-powered sled capable of navigating over snow date back to the early years of the twentieth century, but the first snowmobiles were cumbersome and inefficient. Historian John Warren has uncovered evidence that at least one primitive snowmobile was in use near Willsboro by 1925. But it wasn't until around 1960 that what we would recognize as a direct ancestor of the contemporary snowmobile appeared in the Adirondacks, quickly attracting enthusiastic users. By the late '60s, snowmobile derbies and cash prizes were luring current owners and potential buyers to Lake Placid, Tupper Lake, Old Forge, and other towns throughout the Park. Local snowmobile clubs promoted their use, planned group rides, and groomed trails.

When I was living in Long Lake in the early 1970s, a popular weekend-night event was a poker run. Snowmobilers would ride from bar to bar—Long Lake had five at that time—and collect a playing card at each one. At the end of the night the rider with the best poker hand won a bottle of liquor. On wintry Saturday nights, thirty or more might roar past my house, which sat on a back road linking two of the Long Lake bars.

From the beginning, snowmobiles and their riders were seen by local business interests as the (or at least *a*) solution to the moribund winter economy. Marinas, where boat sales and rentals generated summer income, adopted snowmobiles to keep their businesses open after summer tourists had gone home. Restaurants, bars, and motels that used to close at the end of deer season quickly exploited the option to stay open through the end of winter. Snowmobile operators saw little difference between private and state land and began zipping through the Forest Preserve on foot trails and, especially, old logging roads where they crisscrossed state land acquired from paper and lumber companies. The conservation bureaucracy initially had no regulations for snowmobile use on state land.[1]

But environmentalists were concerned. Growing apprehensions about the noise and the very idea of mechanized recreation in the Forest Preserve were key elements in the story of the establishment of the Adirondack Park Agency in 1971. Environmentalists suspected that the state environmental bureaucracy did not have the capacity or even the will to address the appropriateness of snowmobiles on state land and to prevent them from disrupting what made the Forest Preserve a sacred space. When Brad Edmondson was conducting the interviews for *A Wild Idea*, he recorded Almy Coggeshall and David Newhouse, both of whom were intensely active in Forest Preserve issues beginning in the 1960s; they spoke of how the environmental community understood the rising popularity of snowmobiles to be a threat to the sanctity of wilderness in the Adirondacks. They saw snowmobiles as a looming problem with a lobby powerful enough to prevent the state legislature from regulating them. Newhouse, working with the Adirondack Mountain Club, and Paul Schaefer, at the Association for the Protection of the Adirondacks, among others, began to contemplate the need for litigation if the State either declined to address the problem or, worse, actively promoted the incursion of snowmobiles and their trails into the Forest Preserve.[2]

These were the circumstances when the Temporary Study Commission on the Future of the Adirondacks (which proposed the Adirondack Park Agency) began its work in 1968. The cluster of wilderness defenders around Schaefer, Coggeshall, and Newhouse, including Bill Verner, saw it as the means for forcing the state bureaucracy to recognize the incompatibility of snowmobiles and wilderness. The use of snowmobiles on private land, of course, was another matter altogether, and environmentalists hoped that they would largely be confined to private land. When Peter Paine, a member of the Temporary Study Commission and a charter member of the Park Agency, began drafting the State Land Master Plan, of which he was the primary author, he saw regulating snowmobiles on state land as one of his main objectives.[3] The plan that Paine put together organized the Forest Preserve into distinct classifications, chiefly Wilderness and Wild Forest. Wilderness Areas were the most strictly regulated, Wild Forest less so. It was approved by Governor Rockefeller in 1972 and, with respect to snowmobiles, aimed for a compromise: snowmobiles would not be entirely prohibited but were to be regulated and permitted on roughly half the Forest Preserve, the lands designated as Wild Forest.[4] Along with all forms of mechanized recreation, they would be forbidden in the Wilderness Areas. Whatever the SLMP prescribed, of course, Article 14 remained the fundamental guidance. No construction or maintenance of a snowmobile trail could go beyond what was permitted by the article itself or in the interpretation thereof provided in *MacDonald*.

The 1970s and 1980s saw snowmobiles become a major factor in winter recreation in the Adirondacks, and by the 1990s and the years of the George Pataki administration (in office 1995–2006), the State and the central Adirondack towns were discussing wider trails both for facilitating high-speed, town-to-town sledding and for the gasoline-powered groomers needed to maintain the trails. Peter Bauer, at that time executive director of the Residents' Committee to Protect the Adirondacks, repeatedly warned the DEC that these proposed trails, as they were characterized by the DEC, appeared to violate the constitution. And Dave Gibson did the same for the Association for the Protection of the Adirondacks.[5]

In 2006 the DEC released a "conceptual snowmobile plan," with details about links joining central Adirondack towns.[6] These became known as "Community Connector Trails." Jointly with the Park Agency, the DEC issued a further document, "Management

Guidance: Snowmobile Trail Siting, Construction and Maintenance on Forest Preserve Land in the Adirondack Park," three years later to outline how these trails would be built, how wide they would be, and how they would be maintained. Newly formed Protect the Adirondacks! examined these documents and concluded that the constitutional issues had not been adequately addressed, that, in fact, the trails proposed by the DEC could be built as described only if the constitution, *MacDonald*, and *Balsam Lake* were ignored. Meanwhile, the Andrew Cuomo administration (in office 2011–21) actively promoted Adirondack snowmobiling, while the governor participated in carefully orchestrated photo ops of himself and his family snowmobiling on Adirondack trails.[7]

In addition to the questionable constitutionality of the trails themselves, a further factor in this affair lies in an important series of land deals. In 2007, the Adirondack Chapter of the Nature Conservancy announced that it had purchased 161,000 acres—in scattered parcels, some huge—that had belonged for generations to Finch, Pruyn and Company of Glens Falls. These included lands that Adirondack conservationists had hoped for many years would someday be major enhancements of the Forest Preserve and its protected wilderness: Boreas Ponds, the Essex Chain of Lakes, OK Slip Falls, critical and remote Hudson River shoreline, and thousands upon thousands of acres of peaks, wetlands, lakes, rivers, and forest. The Conservancy developed a mix of plans for all this land: some would remain in forest management, some would continue to be leased to rod-and-gun clubs, but key parcels would eventually be bought by the State and added to the Forest Preserve.[8]

All of these outcomes involved delicate maneuverings on the part of the Conservancy, which needed, among other things, to conduct intense on-the-ground assessments of the ecological and historic attributes of its newly acquired holdings and to cover huge expenses that included interest on borrowed money and voluntary tax payments to local governments and school districts. All this took five years, but by 2012, it had agreed with the State that some 69,000 acres would be bought, for just under fifty million dollars, for the Forest Preserve.[9]

The primary pot of money the state of New York had to draw from for this massive deal—really a series of deals, with separate contracts discussed and then agreed upon for each chunk of land—was the Environmental Protection Fund, created by the legislature in 1993 for addressing major environmental problems and, as in this case, for seizing opportunities for acquisition of critical lands.

One of the enduring local myths in the Adirondack Park is the conviction that when the state buys previously privately owned land and adds it to the Forest Preserve and its protected wilderness, this amounts to an economic loss for year-round residents: the land won't be logged or developed, and thus, the story goes, there will be a loss of jobs. There is no evidence to support this locally held certainty (which conveniently ignores both the state payment of real property taxes on land requiring no services and all the money spent in the Adirondacks every year by tourists and seasonal residents drawn to the region precisely because of its protected open space). The relentless persistence of this fear nearly always muddies the waters whenever any opportunity to expand the Forest Preserve arises.

So when the legislature worked out the details of the state's Environmental Protection Fund in 1993, it offered local towns a veto power when the fund was drawn on for some (but not all) land to be acquired for the Forest Preserve.[10] Local towns did not have absolute veto power over all the Finch lands that the Conservancy wanted to sell and that the State wanted to buy, but Governor Cuomo and the DEC wanted key town supervisors to approve the whole package.[11] To secure their consent, the State signaled a reinvigorated interest in building a network of Community Connector Trails. These trails would ostensibly provide substantial economic benefits for Long Lake, Indian Lake, Minerva, Newcomb, and North Hudson. Whether the State "promised" these trails or merely "committed" to them is unclear.

In 2012 the DEC began construction of the first of these trails, beginning on Route 28 near the Seventh Lake state boat launch. Protect the Adirondacks invoked Section 5 of Article 14 to secure standing for initiating litigation and in early 2013 sued the State, alleging "that the construction of the trails is impermissible because it required cutting and destruction of a substantial amount of timber, would create an 'artificial man-made setting' in the Forest Preserve and was inconsistent with the Preserve's wild forest nature."[12] After multiple rounds of briefs, the official proceedings began in the Supreme Court in 2017 in Albany (for litigation like this in New York the Supreme Court is the entry-level court).[13] This was the trial in which I testified as an expert witness, mainly about my understanding of what the word "timber" meant when the forever-wild provision was incorporated into the Constitution of 1894. Protect's attorneys John Caffry and Claudia Braymer and its witnesses argued that the Community Connector Trails violated the constitution, while the New York attorney general's office, representing the DEC and the APA, argued the opposite.

This was my first of two experiences as an expert witness, and I was filled with apprehension about how it would go.[14] John Caffry rehearsed the whole thing with me, more than once, but I still felt after my testimony that I could have done a better job. In addition to ordinary performance anxiety, my main problem was the difference between how lawyers and historians define evidence. To me an article in the *New York Times* that I find online is just that: an article from a reputable newspaper that I can print for my files and then quote or refer to in an article or book. But in a courtroom nothing like this can be introduced unless it has been certified by, for example, someone at the New York State Library. How one would go about faking an old newspaper at the *Times* website is beyond me, but that is the sort of thing over which lawyers and judges haggle. And it prevented the introduction of a piece of evidence that supported our case. This was the 1915 letter from Louis Marshall to the *New York Times* published right before the vote that year on a new constitution (discussed in the chapter about the 1915 Constitutional Convention).

At the trial, I was the first witness, and my testimony largely depended on the findings I introduced in this book's chapter about the 1894 convention and the contemporary understanding of the word "timber." After establishing my credentials, Caffry methodically led me through this labyrinthine topic. We outlined the dictionary definitions of "timber" in the 1890s and used them along with examples from contemporary usage to show that that word often meant far more than just trees large enough to furnish lumber suitable for construction. While that meaning was certainly one possibility at the time the constitutional provision was drafted, we showed, effectively it turned out, that it wasn't the only one. The state was insisting that the word meant *only* big trees, and we showed that this was not the case.

Our subsequent witnesses testified for several days about the amount of tree cutting and other disruptions actually occurring on those parts of the Community Connector Trails that had already been laid out and constructed. This involved long testimonies about on-the-ground research, with photographs and statistics about thousands of stumps, small and not so small. The point here was to show that the amount of cutting went far beyond the metrics established in *MacDonald* and *Balsam Lake*.

A further argument advanced by Protect was that the connector trails were essentially roads and thus violated the provision of Article 14 guaranteeing that the Forest Preserve be maintained as "wild forest lands." To the contrary, the State repeatedly argued that the new snowmobile corridors were more like foot trails than roads. How a transportation route—the construction of which involved heavy machinery, sturdy bridges, elaborate bench cuts, and, on curves, a width of up to twenty feet—can be understood as a foot path seemed absurd to Protect. And a majority of the justices on the Court of Appeals eventually saw this matter the same way.

In the initial decision, Supreme Court Judge Gerald W. Connolly largely accepted our argument about the meaning of timber. Judge Connolly acknowledged that the "terms 'tree and timber'" had been used "interchangeably" and that the use of the word "'timber' in the constitutional provision at issue herein refers to all trees, the definition of which includes any independent growth of a species that is biologically identified as a tree, no matter the size." Nonetheless, relying on his assessment of the decisions in *MacDonald* and *Balsam Lake*, he ruled that the amount of cutting that the state was undertaking did not violate the parameters laid out in those cases, and he further declared that

Protect's argument that the new trails were essentially roads and thus compromised the "wild forest lands" provision was not persuasive.[15]

Protect appealed, and the case moved to the Appellate Division, Third Department, the second step (of three) in the judicial sequence for cases like this one. Once a case is in the chute for appeals, testimony from witnesses and additional evidence are not solicited or permitted. The lawyers for each side submit briefs and counter briefs, all based on evidence already in the record, and then present oral arguments to the panel of five judges. Caffry and Braymer were again making the case for Protect. The nature of the contest continued to focus on the key question of constitutional intent. What does Article 14 mean? What did its authors intend? How have those answers to those questions been, at least partly, answered or perhaps suggested by the decisions in *MacDonald* and *Balsam Lake*? It's a rather baroque ritual involving close reading of occasionally antique language, but it's how we operate in the United States. And it generally works—cumbersome but ultimately productive.

In its decision, the Appellate justices followed the argument, established by Protect at the outset of the whole process and employed as an organizational device by Judge Connolly, that the Community Connector Trails had to be examined on two grounds, i.e., with reference to two key elements of Article 14. First, did construction of the trails violate the mandate that the Forest Preserve be kept "forever kept as wild forest lands," and second, did that construction require an unconstitutional destruction of the timber? On the first of these, the court found the trails to be constitutional: the trails did not of themselves offer a threat to what had traditionally been understood—always within the parameters of *MacDonald* and *Balsam Lake*—to the constitutional requirement that the Forest Preserve be "kept as wild forest lands."

But on the second criterion, destruction of timber, the court ruled that the trails in fact presented a violation of Article 14. Much to my delight, the decision cited my testimony about the broader meaning of "timber" in reaching this decision. "We agree with Supreme Court's determination, based on the expert historian's testimony as well as other evidence, that the use of the word 'timber' in the constitutional provision at issue is not limited to marketable logs or wood products, but refers to all trees, regardless of size."[16]

It was a bifurcated decision: the State won on the first argument, and Protect won on the second. Since the loss on the second issue meant that the on-the-ground work of the State in constructing these trails had to stop, the State of course appealed that part of the decision. And since this meant Protect was heading back to court anyway, Protect opted to appeal the first part of the ruling, the whole issue of whether the connector trails by their very nature violated the meaning of "shall be forever kept as wild forest lands." This was critical: neither *MacDonald* nor *Balsam Lake* had asked the courts to rule on whether a state action could be understood as a violation of the principle of "wild forest lands," in other words as a violation of the State's constitutional obligation to protect wilderness.

At this point, I guessed that the Court of Appeals would repeat the ruling of the Appellate Division: no violation of forever wild, but too much tree cutting. (I was wrong.) As with the appeal to the Appellate Division, the next step in this elaborate ritual involved briefs and counter briefs, with no additional evidence or witnesses, followed by oral arguments from attorneys before the Court. John Caffry argued for Protect.

Before the oral arguments, the whole process was also attracting the attention of other parties. By the time all papers had been duly submitted to the Court of Appeals, amicus briefs—arguments from organizations not suing or responding in the original filing— were also submitted. These supported the case of either Protect or the State. The Adirondack Mountain Club (ADK), the Open Space Institute (OSI), the Nature Conservancy (TNC), the Association of Adirondack Towns and Villages, and the Empire State Forest Products Association entered the fray on the side of the State. And the Sierra Club, Adirondack Wild, and the Adirondack Council supported Protect. The fact that ADK, TNC, and OSI entered briefs in support of the State's position seemed surprising. ADK argued that if Protect prevailed, foot trail maintenance and construction would be threatened, while TNC and OSI, because they had participated in the negotiations with the towns during the lengthy process of state purchase of the Finch lands, argued that the State had committed to the connector trails and that a win for Protect would render that commitment impossible to fulfill.[17] Whether or not the State had in fact promised the trails is of course irrelevant to the question of the trails' constitutionality: the State, or its bureaucratic representatives, cannot rewrite the constitution by simply declaring its intention to do something.

Whether the amicus briefs had any impact on the Court of Appeals is unclear. I assume the Justices read them, but were they influenced by the quite predictable fact that the Association of Adirondack Towns and Villages wanted the connector trails built and used or that the Sierra Club did not? The Court may have been surprised by the perhaps counterintuitive positions taken by ADK, TNC, and OSI, but did they make any difference? The final decision suggests not.

25

The Decision in *Protect*

Meanwhile, the Court of Appeals was preparing its decision, which turned out to be a resounding victory for Protect and, more important, for wilderness and the Forest Preserve. In May 2021, the Court of Appeals upheld the Appellate Division's ruling with respect to the meaning of "timber" and ruled that construction of the trails was in fact unconstitutional.[1] And, in a historically significant part of their decision, the court further ruled that the trails violated the crucial provision that the Forest Preserve "shall be forever kept as wild forest lands." If the people wanted these trails built, they would need to pass an amendment to the constitution. The intent of the delegates of 1894 could be reasonably assessed, and if New Yorkers wanted some different understanding of the Forest Preserve, amending the constitution was the only route. It is not within the authority of the DEC to interpret the constitution to suit its priorities.

The Court of Appeals explicitly rejected the assertion of the Appellate Division that Article 14 had to be read as "bifurcated," with a caesura occurring somewhere between "wild forest lands" and the prohibition of the destruction of timber: "All members of the Court agree that the constitutional protection is unitary" and that any destruction of timber is a "violation of the 'forever wild' clause, because that prohibition was a means to the ultimate objective of protecting the forest as wilderness." The court's insistence that the "ultimate objective of protecting the forest as wilderness" was what the 1894 convention intended to accomplish is a dramatic and striking claim, and short of amending the constitution, it changes forever how the Forest Preserve can be managed. Equally important, it clarifies how the

Forest Preserve is to be understood. The evidence for this conclusion is clearly in the record of the constitutional convention, as I have tried to show, but I was not expecting the court to see it this way, especially not so adamantly expressed.

Much of the decision is devoted to statistics of trees to be cut, rocks to be moved, acres to be cleared, and the difference between a foot trail and connector trail. All of this is vital to the final decision because of the arguments presented by both sides in the lower courts, but the nut of the decision, the part that breathes new life into the very idea of wilderness in the Adirondacks, is the dismissal of the narrow notion that only tree counting can lead to a finding of unconstitutionality. The constitution, this decision insists, says that the Forest Preserve must "forever kept as wild forest lands," and from the date of this decision until an amendment or until another court revises it, that is an expression that we must adhere to rigorously. I do not think that the drafters of the 1885 Forest Preserve law, where that expression became statutory and where the delegates of 1894 found it for inscription in the constitution, had a twenty-first-century understanding of wilderness in mind. But that understanding begins to appear and is rigorously promoted in the debates of 1894. Constitutions are living documents, and it is the role of the courts to keep them alive. Politicians and state agencies cannot give us a definitive assessment of what the constitution means, either when it was written or in the present. But courts can and do. And the Court of Appeals, after extensive perusal of the record and after hearing extensive arguments in this case, has told us how it understands Article 14 and its protection of New York's remaining wilderness.

The court also rejected the State's argument that the trails in question were legitimate because they "enhance access to the Preserve and provide a variety of recreational opportunities" and even worked Robert Marshall's famous "The Problem of the Wilderness" into their decision. As the court had argued in *MacDonald*, access is indeed important, but if a specific form of access violates the constitution, it cannot be introduced without an amendment, regardless of how appealing it may appear. The court opined more than once that an amendment could be passed to legitimate the connector trails, but without one, Article 14 means what the court has told us it means: Article 14 can be understood only as a solid, non-negotiable protection for wilderness in the New York State Forest Preserve.

Among staff and board members at Protect, there was jubilation. We knew that the team of John Caffry, Claudia Braymer, and Peter Bauer had performed heroically and that without their dedication and invaluable labor and energy this historic moment would never have occurred. In some other corners of the Adirondacks, the reaction was not so positive. Local government officials, who had negotiated carefully with the State about additions to the Forest Preserve from the former Finch, Pruyn lands, thought they had been promised the Community Connector Trails. Supervisors from Indian Lake, Minerva, Newcomb, North Hudson, and Long Lake could have stalled the state purchase of some 69,000 acres of Finch lands, but they went along once they thought they had a promise from the DEC that the connector trails would be built. When the decision from the Court of Appeals was released, they understandably believed that they had been lied to. Exactly what the State had told them is not a matter of public record, and former DEC Commissioner Joe Martens said in a Zoom conference in June 2021 that it was inaccurate to use the word "promise" to define the State's position.[2] Perhaps, but there is no doubt that local officials sincerely believed that a promise had been made. Dominic Jacangelo, executive director of the New York State Snowmobile Association, told a reporter from the *Explorer,* "Every town supervisor I talked to was promised things, which obviously now cannot be fulfilled. Mainly they were promised connecting trails."[3]

The reaction from local government reflects the endless contest over wilderness. To Protect (and me) wilderness is real, guarded by the constitution and worth energy, time, and resources to preserve. To the supervisors of the towns that would have been linked by these trails, wilderness is a fictional construct, imposed on them by outsiders indifferent to the economic needs of year-round Adirondack residents and businesses.

Notwithstanding Protect's glee upon reading the decision and its hope that the matter was settled, the DEC initially dodged any discussion of changing its plans. It took two years of negotiating, foot dragging, and threats of further litigation, but finally in October of 2023, a decade after the beginning of this historic litigation, Judge Connolly, who had heard the case and ruled in the Albany Supreme Court, reaffirmed the decision from the Court of Appeals and awarded some minimal costs to Protect. This seemed to be the final word: the trails were unconstitutional.[4] But in the fall of 2024, the State backtracked and floated new guidelines for snowmobile construction that clearly ran counter to the spirit of the decision from the Court of Appeals. When it comes to wilderness and how officials at the state's conservation bureaucracy choose to honor its protection—or not to do so—the battle is never over.[5]

26

Wilderness in the Sewards

One July day in the late 1980s, I set off, solo, for a few nights in the Sewards, my sanctum sanctorum. I started at Plumley's and plodaingly worked myself into the proper frame of mind on the trail to Shattuck's Clearing. Hiking this stretch of trail as a teenager, eager to get to Cold River, I used to want to get it over with as quickly as possible—it's relatively flat and unexciting to the adolescent mind. But for many years, I have found this familiar trail, with its mix of hardwoods (some huge), pines, and wetlands a reassuring and friendly warmup for what's to come. George Marshall, recalling the dusty roads that he, his brother, and Herb Clark routinely trekked from the Lake Placid train depot to the northern entries to the High Peaks, valued how their route provided a "transition, physically and psychologically, between the twentieth-century world and the primeval." For him, it was an important liminal space between home and "the mood of forest and mountain."[1]

Near Shattuck's I left the Northville-Placid Trail and hiked north on the Calkins Creek horse trail—probably more like the road mentioned by George Marshall—and then bushwhacked up the west side of the Seward Range, following Boulder Brook and the elusive traces of an old tote road.[2] In the twenty-first century a few other hikers have used this route, but back then it was all mine (or, like Bob Marshall in the Brooks Range, I thought it was). It was my quintessential, always paradoxical, occasionally illogical Adirondack wilderness experience.

I especially wanted to take another look at a magnificent white pine in the lower reaches of Boulder Brook, which I had first seen with Bill Verner about fifteen years before. This tree, I soon discovered, had been blown down and was beginning the slow process of its disintegration, returning its billions of cells to the soil, as any properly respected wilderness tree should. The brook is a splendidly typical high-peaks stream, with crystalline water swirling past or over rounded anorthosite boulders: gentle and meandering at lower elevations, loud and rapid up where it's steep, with noisy falls dropping into densely shaded pools. I spent my first night next to the brook, sleeping well in my tiny tent. Not a trace of another human being nearby—except for the tote road I'd been following, the last vestige of which disappeared, at least to my eye, long before I set up camp for the night.

The next day, leaving pack and tent behind, I headed for the blowdown, cripple bush, and krummholz toward the top of the ridge. At one point, precariously balanced on the deteriorating trunk of a tree knocked down by the 1950 storm, forcing my way through the prickly little spruces, and warily eyeing the moss-covered rocks a couple of feet below me, I pondered my situation. I'm a day's hike from the nearest phone; I'm by myself; if I break a leg now, they won't find me for days. This was before the age of cell phones and pocket devices that can send an SOS signal to a satellite (both of which I now carry). I was too tired and sweaty to contemplate this for long, though, and continued thrashing my way up to the ridge between Emmons and Donaldson. Once there, I followed the herd path to the Donaldson summit, where I feasted on the view and the isolation for a spell before retracing my steps and heading back to my snug campsite. I saw not another soul all day.

The reason that this place in northern New York can be routinely referred to as wilderness follows from the extraordinary provision of the state constitution, now (but not then) clarified by the decision in *Protect*. There are a number of distinct but often overlapping ideas that many of us have in mind when we think about wilderness. At times we emphasize the experience that we believe is available in wilderness, an experience that is obviously different from what we ordinarily know in our urban, suburban, or even rural lives. We may be thinking about a certain kind of wildlife habitat, one where species that depend on wild country thrive. In the contemporary United States, we're probably thinking about a precisely bounded place, administered by a government agency, where certain human activities may have occurred in the past but are now proscribed. We may be distinctly aware that bureaucratic, administrative classification of wilderness is a compromise, an artifice of the very culture we have slipped away from.

In a world characterized by frantic, relentless, often useless, apparently uncontrollable change, moreover, we're probably also looking for some sense of permanence. The "preserve" in "preservation" indicates, among many other things, our yearning for a place that stays the same, a place where change does not happen so fast and so unpredictably as it increasingly does in the social, commercial, cultural world where we spend most of our time and where we often feel dislocated and alienated. We want a place that looks like it used to, or at least like what we think it used to. When we find it, as I did that day, we feel spiritually moved and psychologically refreshed.

This is what one of the best of twentieth-century Adirondack writers, William Chapman White, had in mind when he penned some lines that the many fans of wilderness in the Adirondacks often like to quote: "As a man tramps the woods to the lake, he knows he will find pines and lilies, blue heron and golden shiners, shadows on the rocks and the glint of light on the wavelets, just as they were in the summer of 1354, as they will be in 2054 and beyond. He can be part of a time that was and time yet to come." And, White added, this fortunate state of affairs will obtain, "so long as the forest preserve is in the constitution."[3] But White was writing without the benefit of William Cronon or several decades of challenge to such notions of permanence and stasis. If we are searching for this sense of permanence, we are doing just what Cronon warned against: removing the land from history. And this spot, just like any spot, had plenty of history, both anthropogenic and natural.

On Donaldson, I had solitude—to the point of potential peril. I had beautiful and ostensibly wild country all around me—nothing but a dense, lush forest and lonely mountains in every direction. Instead of the blurry cacophony of air conditioners, rock music, and vehicles, all I could hear was the occasional distinctive trill of a white-throated sparrow and the gentle vibrations of the breeze in the balsam firs. The sky was heartbreakingly blue, not the foul yellow smudge typical of the summer air in most of urban America. I was sitting on the top of a peak with no maintained trail, considered by many Adirondack hikers to be one of the most remote spots in the Park, maybe in the eastern United States. It was easy to dream that I was repeating the experience of Verplanck Colvin and Alvah Dunning, who together made the first known ascent of Donaldson in 1870.[4]

But there was that tote road. And what about the herd path along the ridge, stretching in one direction to Emmons and in the other to Seward and then down to the Ward Brook fire road? What about the depredations of the Santa Clara Lumber Company, which had worked over the Sewards beginning in the mid-1890s? They built tote roads and bridges, cut all the pulpwood they could sell, and set up winter camps, where they left assorted debris.

When Colvin and Dunning first reached the summit of Seward on an icy October morning in 1870, the Seward range showed no evidence of any prior human impact. In his account of the ascent, Colvin referred repeatedly to the slow progress he and Dunning made through the dense forest, often "compelled . . . to clamber upon hands and knees."[5] Colvin and Dunning enjoyed the kind of experience that stirs the envy of today's climbers: this was raw, ostensibly unexplored and untouched nature. But during Colvin's lifetime, the conditions that provided the wilderness experience for them and for any others that may have followed them were hacked away by the axes and saws of the Santa Clara Lumber Company.

When Robert and George Marshall and Herb Clark climbed the Sewards shortly after Santa Clara finished logging, the forest was beginning its slow recovery. Their experience was remarkably different both from Colvin's and from that of most of their Forty-Sixer descendants. In *The High Peaks of the Adirondacks*, Robert Marshall described the hike: "We climbed the mountain from the Cold River side, following lumber roads to the edge of the slash and found it an easy climb." He added, "Lumbering operations scar the view toward Cold River."[6] Veteran Adirondack climber Jim Goodwin, who made the ascent shortly after the Marshalls, found that even above the tote roads, "climbing was quite easy as the slopes were then open through park-like spruce and balsam forests."[7]

The State bought the Sewards from Santa Clara in the 1920s and added the southern half of Township 27 of Tract One, Macomb's Purchase, to the constitutionally protected Forest Preserve.[8] There would be no more logging on the slopes of the Seward Range. For thirty years, the forest slowly matured, subject to no human intrusion (other than the occasional aspiring Forty-Sixer, and there weren't many of those back then) and to the ordinarily patient processes of nature; on the upper slopes, where Santa Clara loggers had cleared out the spruce, fast-growing, shade-intolerant white birch occupied much of the ground.

But the violent windstorm of November 1950, the great blowdown, showed that nature can affect sudden, dramatic change to the landscape. The trees on the upper slopes of the Seward Range were thrown down with devastating force. The forest through which the Marshalls and Jim Goodwin had strolled became a tangle of fallen trunks and rotting, almost impenetrable brush. Because of the fire hazard in this vast pile of tinder, the Sewards along with all of the Cold River Country were off-bounds to hikers for five years. When climbers were allowed back in, they found a forest quite different from what they remembered—a forest, in fact, that was more like the one encountered by Colvin and Dunning. By now scrub balsam and spruce were beginning to grow up through the blown-down birches, and navigating through, over, under, and around this mess became a uniquely Adirondack challenge.

In other words, along with the landscape, the experience one encountered in the Sewards was changing. It changed during Colvin's life because of human activity; it changed in the '30s and '40s as the cut-over forest began to recover; it changed in a relatively sudden moment in 1950 when a ferocious wind swept out of the East; and it continued to change. When I first climbed the Sewards in the 1960s, the blowdown was still more or less sound. You could walk on the trunks, while fighting your way through the balsam and spruce. Now the blowdown is completely rotted away, and every one of the original "trailless" peaks has an easily followed trail. There's a huge slide on the east slope of Emmons. A cloudburst in the 1980s dropped a few tons of water on some loose soil near the Emmons summit, and the result was a dramatic scar and a new climbing route. I've used it to approach the Emmons summit.

So are the Sewards a wilderness? Is the experience of climbing in the Sewards a wilderness experience? The forest is certainly not "virgin." The Santa Clara Lumber Company saw to that. The range holds all sorts of evidence of human interference in the landscape: traces of tote roads, cast-iron relics abandoned at logging camps, herd paths linking the summits. Isolation? The path between Seward and Donaldson is certainly less populated than downtown Tupper Lake or, for that matter, the trails from the Johns Brook Valley to Marcy, but in July or August your chances these days of encountering another climber or two are pretty good.

In terms of solitude and pristineness, the Sewards—like most of the Forest Preserve—fall into a rather ambiguous zone. Yet the State of New York declares that the Seward Range, along with over a quarter of a million acres of nearby mountains and forest, is part of what it officially calls the High Peaks Wilderness Area—so designated because the land, the State Land Master Plan avers, is "an area where the earth and its community of life are untrammeled by man." The SLMP goes on to insist that wilderness is "an area of undeveloped forest preserve land retaining its primeval character and influence."

I sense more confusion here. If this is a "community of life . . . untrammeled by man," then where are the eastern timber wolves that we know inhabited the Adirondacks (barely) in the age of Colvin and Dunning? Where are the cougars, whose decline and impending extirpation Colvin himself lamented just as the Santa Clara loggers were laying out their tote roads? Where are the giant white pines, hints of whose monumental presence can be detected in a few relics down in the lower reaches of Boulder Brook?

Our culture's need for wilderness—even though we're not sure what it is—begins with a sense of regret of something lost in the modern, industrialized world. We want to protect and experience places where the ravages of urbanization, overpopulation, and pervasive environmental degradation don't leap in front of us at every turn. Yet even here in the Adirondacks, even in the areas we have designated as wilderness, we have, if we look for it, signs of the ravaging of the American landscape by industrial capitalism. For what were the logging operations of the Santa Clara Lumber Company but one more incident in the massive despoliation of the American land, all in the name of profits and progress, that accelerated after the Civil War and continues today?

But despite all these quibbles, I know that there's something special in the Sewards and that we can thank the New York State Constitution for it. The forest is largely second growth, the cougars are gone (though that doesn't mean they will never come back), the woods are full of little reminders of the people who've been there before. But the experience of negotiating blowdown and cripple bush, of eating sardines and pita bread and drinking warm water from a plastic canteen, of staring into Ouluska Pass from Donaldson, is nonetheless spiritually moving. Sitting on the summit and peering over at the looming mass of Seward, down into the broad, green swath of Cold River Flow—itself a relic of a long-gone logging dam—or over to the bluish, slightly shimmering ridges of the eastern High Peaks, we are aware of both change and at least the comforting illusion of permanence.

We also know that we have the anxieties of 1894 to thank for this moment. Because the water level in the Hudson seemed dangerously low in that dry summer, the Seward Range is in a state of gradual repair. The Santa Clara loggers did not do irreparable damage, no more than the great blowdown did. Because the Seward landscape is not pure, because the evidence of human-caused change is omnipresent, but because the land at the same time is so obviously not a landfill, an industrial ash heap, or a teeming shopping mall, it is both of the world we know and a reminder of the world we've lost. It's a place where people have wreaked havoc but which to a large extent has recovered. It's also a place where nature has precipitated landscape change quite as dramatic as that accomplished by people: the blowdown of 1950 and the slide on Emmons, in their simultaneous similarity to and their obvious difference from late nineteenth-century logging, remind us that whatever wilderness might be, it's not permanence. Nature, even when left alone—if that's possible—doesn't stand still. It is dynamic, cyclical, full of apparent disasters just waiting to happen. One of the things that wilderness is and that the constitution protects it for is to be a place where natural changes can occur and where a landscape scarred by human abuse can heal.

The idea of wilderness in the Adirondacks is ambiguous, elusive, paradoxical, and pretty much impossible to defend if someone just doesn't want to buy it. When a friend wants to point out that the Adirondacks, like nearly all of North America, were the home, or at least a familiar seasonal hunting area and occasional locus for crop cultivation, for the people who lived around here before any Europeans showed up, that the Adirondacks have been the site of often intense resource extraction, that the Adirondacks are now subject to the anthropogenic alteration of climate and weather, and that the Adirondacks for all these reasons manifest many marks of human impact, I have to agree. To that person, convinced that only land that has never been logged or otherwise exploited can ever be considered wilderness, my position makes no sense. But it works for me, and apparently for thousands of others.

Or if another friend argues that the protection of wilderness is simply a pandering on the part of our governments, federal or state, to the needs of a few, I recall a brief article in a Forest Service newsletter, "The Wilderness as a Minority Right," published by Robert Marshall in 1928.[9] Marshall was at the beginning of the campaign that would finally conclude, long after his death, with the 1964 Wilderness Act and the classification of certain federal lands as protected wilderness.[10] The initial efforts from him and others to convince the federal conservation bureaucracy to set aside wilderness lands immediately ran into resistance. Their foes argued that protection of large parts of the federal domain and setting them aside for the use of the then relatively small number of wilderness enthusiasts was cloistering publicly owned resources merely to satisfy the desires of an elite minority, that those lands could otherwise be logged, mined, or grazed on to the benefit of all.

Responding to this position, Marshall pointed out that the number of wilderness enthusiasts was growing. More important, he maintained, there are many ways we devote resources to satisfy a minority. If wilderness is an elitist concern, then so, too, are museums, concert halls, universities, and swimming pools, all of which are supported by public funds but are not universally enjoyed. In the twenty-first century, he might have added the enormous sports stadiums so lavishly subsidized by tax dollars in cities around the country. Wilderness enthusiasts have rights, too. And this minority is asking for the use of only "a minor portion of America's vast forest area for the nourishment of this peculiar appetite." One could say the same about designated wilderness in the Adirondacks: it constitutes about one sixth of the Adirondack Park and one thirtieth of New York State. "The necessity of getting away from the stifling artificiality of civilization," he continued, "cannot be explained to those who have never apprehended the passion for the wilderness, which is just as genuine as the more conventional yearnings for love and beauty."

Article 14, enjoying greater clarity and force with the decision in *Protect*, protects the Sewards and all the wilderness in the Adirondacks and keeps the forces of industrialism at bay. It can't be justified to everyone. Nor can it stabilize the temperature—of which more in a subsequent chapter—but like the idea of constitutionalism itself, it's the best we have. And it's worth defending when the state bureaucracy wants it diminished. Just as Robert Marshall was doing in the Brooks Range, I was in a sense performing my wilderness experience in the Sewards. But—and this is important—that is not to say I was fooling myself, or merely pretending. The constitution and history come together and have provided us a holy place, and it is no act of delusion to call it wilderness. A trek in the Sewards, or in the Brooks Range, can be both performative and transformative.

27

Whose Wilderness?

In the late nineteenth century, White men ran New York and created the Forest Preserve. That dynamic is changing. In the decades ahead, how will New York citizens ethnically different from the delegates of 1894 understand the protection of the Forest Preserve? If wilderness has historically been both a construction and a treasure of—mostly—White people, especially comfortably well-off White men, how will it fare in a state no longer as White as it once was and becoming increasingly less so? The existence of this wilderness, it should be recalled, involves tax payments made by the State to Adirondack local governments and school districts. Will delegates from, say, the Bronx who may feel unwelcome in that wilderness be willing to continue dedicating their tax dollars to the support of small towns and underpopulated schools in the Adirondacks? It remains to be seen whether an ethnically diverse population will put a high value on the constitutional provision now protecting the Forest Preserve, especially since non-White New Yorkers have all too frequently felt unwelcome in the Adirondacks.[1]

The notion that the Adirondack wilderness is the special, proprietary province of a certain sort of privileged White men has deep roots in our history, with its first manifestation surfacing in the response to throngs of ostensibly unprepared sportsmen in the years after the Civil War. In the spring of 1869, the distinguished Boston publisher Fields, Osgood released a book that it hoped would make a buck or two but for which it had only modest hopes. This was *Adventures in the Wilderness; or, Camp-Life in the Adirondacks*, by a Boston clergyman, William Henry Harrison Murray.[2] Much to the delight of publisher and author, the book struck a nerve and became an overnight best-seller. With its appealing tales about Adirondack fishing and hunting and, perhaps more important, with its detailed, easy-to-digest instructions on how to plan and execute a camping trip and a sojourn deep in the wilderness, Murray's book inspired thousands of neophytes to head for the Adirondacks, where they encountered hotels and boardinghouses with no available rooms, swarms of bloodsucking blackflies, a remarkably rainy summer, and widespread contempt for the newbies, forever after labeled as "Murray's Fools." For the rest of his life William H. H. Murray was known as "Adirondack Murray."

Before Murray, the Adirondack wilderness had been enjoyed only by a few cognoscenti. The peak we now know as Mount Marcy wasn't mentioned in print until 1836 and not climbed (so far as anyone knows) until the following summer, when state geologist Ebenezer Emmons led the first recorded ascent and proposed the name "Adirondacks" for the surrounding peaks. A few urban dwellers read about Emmons's explorations and decided that the newly named Adirondacks—difficult to reach and thinly populated—might be a good locale for gentlemanly hunting and fishing. They found what they were looking for, and a few of them published descriptions of their adventures that in turn inspired a few others. Joel Headley's *The Adirondack: or, Life in the Woods* was the best known of our antebellum sporting narratives.

But Headley's readers remained relatively few. The America of the 1840s and 1850s wasn't quite ready for the mass phenomenon we now know as the middle-class family vacation. By 1869, however, the combination of post–Civil War exuberance (in the North), improved steamboat and rail travel, and Murray's accessible, often humorous prose made his book a blockbuster. Where sportsmen in the Headley era were paddled by their guides from the Fulton Chain to the Saranacs and rarely encountered another soul, after 1869 the appeal of the northern wilderness had been definitively announced. Spartan farm dwellings were turned into boardinghouses, which soon became hotels. Where a few local men had worked as guides for Headley and his kind, after Murray's Fools were on the scene, anyone with a boat and a deerhound could demand top dollar to take rookies into the woods, keep them alive for a few weeks, and show them the good fishing holes.

The idea of the Adirondacks as a major tourist destination was born. And with that arose the ineluctable conundrum of wilderness. When people find a place they like—a place that satisfies their spiritual longing for transcendent beauty, soothes their inchoate doubts about the insidious reach of urban industrialism, and offers healthy outdoor exercise—and then tell others about it in glowing, perhaps hyperbolized terms, will those who read these encomia show up in such numbers that the original charm is diminished, even erased altogether? More important, when these new people arrive in the backcountry, will those already there be willing to share their wilderness, or will they circle the wagons and insist that the newcomers simply don't belong there?

The sudden, dramatic appeal of Murray's book was an amazing thing. Bookstores throughout the East ran out of copies as the publisher frantically ordered additional print runs. In July, just as the size of the Murray Rush was becoming apparent, Fields, Osgood issued a special Tourist's Edition with waterproof cover and map. A reporter for the *Boston Daily Advertiser* declared that "Mr. Murray's pen has brought a host of visitors into the Wilderness, such as it has never seen before." It was, noted this anonymous writer, "a multitude which crowds the hotels and clamors for guides and threatens to turn the Wilderness into a Saratoga of fashionable costliness."[3]

That first summer Murray's Fools found bugs in clouds of biblical proportions, day after day of rain, and no available guides. They crowded into the Adirondacks in July and then departed in August, some declaring that Murray was a liar. But initial disappointment did not end the Murray Rush. The following year and ever since, the Adirondacks were truly on the nation's map. Blackflies and rain would not keep the tourists away. While many of those lured to the Adirondacks by Murray's bestseller never left the veranda of whatever hotel or boardinghouse they could find a bed in, many were paddled into the backcountry, where their efforts to bag a deer or net a brook trout were scoffed at by hunters and anglers who had been just as inexperienced a generation earlier but now felt themselves the duly appointed guardians of the wilderness.

Critics insisted that Murray, by luring the inexperienced, had ruined the outdoor pleasures enjoyed by earlier sportsmen, who had been sure that they had the wilderness all to themselves and always would. Behind this was a sense that those who truly appreciated the wilderness were a special breed, that Murray was encouraging the wrong sort to make the trek to the Adirondacks. Thomas Bangs Thorpe, a writer of national repute, who published an account of sporting along the Fulton Chain in the 1850s, insisted that the sportsman who really belonged in the wilderness was possessed of a "highly-cultivated mind which rejoices in the wilds of Nature" and was appalled on seeing "those temples of God's creation profaned by people who have neither skill as sportsmen, nor sentiment or piety enough in their composition, to understand Nature's solitudes." It was Murray's "fashionable twaddle," opined Thorpe, which lured the inept and insensitive to the wilderness.[4] What Thorpe was expressing was the widely shared notion that the wilderness existed for the enjoyment of affluent White men and no one else.

Kate Field, a well-known journalist, wrote in the *New-York Tribune* that "many sportsmen are rampant because their favorite hunting and fishing grounds have been made known to the public." Offended by the elitism expressed by Thorpe and others, Field declared, "If several hundred men think that the life-giving principles of the North Woods was [*sic*] instituted for the benefit of a few guns and rods, they are sadly mistaken." Fields's entrance into this discussion especially outraged Thorpe. One of Murray's key points in *Adventures* was that the Adirondacks had been for too long an exclusively masculine domain. He described proudly how much his wife loved her time in the wilderness and how she enthusiastically accompanied him on his annual expeditions. To Thorpe this was blasphemy: "We do not consider the wild woods a place for fashionable ladies . . . ; they have, unfortunately, in their education, nothing that makes such places appreciated, and no capability for physical exercise. . . . Let the ladies keep out of the woods."[5] Neither Murray nor Thorpe considered the possibility that the men or women who might in the future wish to camp in the wilderness might be some color other than White.

When we see Thorpe move from a perhaps understandable but poorly expressed lament that the northern wilderness could be compromised by too many people to a brazen and vulgar sexism, we should ponder the implications of any complaint about the violation of one's personal retreat by the wrong sort. Ever since Murray's day, the notion that wilderness of yesterday was ideal but that the wilderness of today and tomorrow is compromised, threatened, and even permanently corrupted by crowds of the unprepared and unskilled has been a constant feature of the Adirondack story (and is not, alas, completely absent from my own reminiscences). In and just after Murray's heyday, it coincided with a wave of immigration to the United States from eastern and southern Europe and with WASP anxieties about the dilution of American vigor by foreign genes and cultures. In 1888, an unsigned editorial in *Forest and Stream* lamented, "Between the fish-hog, the railroad, the Italian railroad hand, the night-hunter, the pseudo-sportsman and the like, this grand region is becoming yearly less and less like its old self and a few more years will witness its entire destruction from a sportsman's and nature-lover's point of view."[6] In the Adirondacks such elitism led to the establishment of huge private preserves where aristocratic chums could hunt and fish unmolested by the fumbling hoi polloi. It also led to the antisemitism that spread its insidious toxins to many regional hotels and clubs.[7]

The elitist claim of paradise lost is a nasty undercurrent in Adirondack history, a local manifestation of a theme nearly ubiquitous in American culture. And it shows up, in modern dress, in some of the responses to the crowds in the Eastern High Peaks. Ponder the reader comments whenever this subject surfaces on the *Adirondack Almanack*: you'll see a familiar refrain about how *our* wilderness is being overrun by the unmannered and unskilled.[8] You'll encounter the same theme in comments on the weekly backcountry rescue reports from the DEC: these people don't know how to take care of themselves and don't belong in *our* wilderness. It is a modern manifestation of the elitism that William Cronon found and rightly skewered in the age of Theodore Roosevelt and John Muir.

From Murray's era to our own, the assumption that yesterday was Edenic and tomorrow will introduce the end of everything we cherished has periodically poisoned efforts to make the Forest Preserve, owned by all the people of New York, a treasure available to and potentially valued by everyone. As New York becomes more diverse, as languages other than English are spoken in every New York county, the Adirondack wilderness needs a constituency of everyone. A century and a half ago, Adirondack Murray began the process of democratizing New York's wilderness. And his efforts encountered immediate and often virulent resistance.

Most of the vitriol visited by the likes of Thorpe on the unwelcome interlopers was directed at women of any ethnicity and at recent immigrant groups. At the time, the idea that Black Americans might enjoy the wilderness was not part of the popular discussion. Nor was the idea that there is a history of Black Americans in the Adirondacks. But the last few years have produced powerful evidence that the Adirondack story is not solely the tale of White people that too many of us have promoted. Melissa Otis has developed the story of Native Americans in the Adirondacks, and Daegan Miller, Sally Svenson, and Amy Godine have shown the existence of a rich, complex history of Black Americans in the Adirondacks.[9]

In 1874 Seneca Ray Stoddard, photographer and author of Adirondack guidebooks, claimed to have encountered a "darkey belaboring a pair of absent-minded and almost absent-bodied horses" near North Elba. Stoddard invoked this meeting as an illustration of his certainty, a common one for the day, that the man was out of place, that Black people could never find the Adirondacks appealing. He deployed the racist, faux–African American accent popularized by blackface minstrelsy and had the man declare that Black people found the region too cold and lacked the necessary survival skills. Stoddard had him recount a probably apocryphal tale about a Black hunter who found himself lost in the woods, distrusted his compass to the point that he destroyed it, and froze to death. Stoddard took this convenient anecdote to exemplify something that must have been reassuring to many of his White readers: "It is a well-known fact that some unused to the woods will become so effectually 'turned around' that they will be certain that something is the matter with the compass to make it point wrong, and even distrust the sun itself if it happens to be in a different position from that which they think it *ought* to be."[10] In other words, according to the casual racism of the day, Black people did not belong in the wilderness.

But there are other threads in this story. The example of Robert Marshall reminds us that in the wilderness movement, at least in the twentieth century and later, there has been a counter and anti-elitist element. To begin with, Marshall was Jewish, and he and his family were well aware of the antisemitism that permeated American culture. They were active supporters of civil rights for minorities.[11] Robert Marshall devoted considerable energy to trying to think of ways to provide American children of all colors and backgrounds the opportunity to experience the wilderness the same way he had as a child. He inherited significant assets from his progressive father and was a generous supporter of civil rights and equity for non-White Americans. He understood the fact that up until his time the American wilderness had largely been the province of privileged White people, and he hoped that the Wilderness Society, among others, would implement programs to democratize the wilderness.[12]

But the idea that America's natural wonders belong to all Americans has routinely encountered resistance. In Marshall's day the unwillingness of some White Americans to share the wilderness surfaced tellingly in two national parks in the Southeast, Shenandoah and Great Smoky Mountains, where the Park Service wanted integrated facilities but local governments demanded that campgrounds and their infrastructure be operated according to the Jim Crow laws then powerfully in place throughout the old Confederacy. A battle over jurisdiction raged for most of the 1930s before the federal authorities simply refused to institute segregationist policies.[13]

New York did not have Jim Crow laws written into its statutes, but segregation constituted an undeniable reality in our public spaces. As a web page maintained by the Office of New York State Parks, Recreation and Historic Preservation notes,

> Vacationing in New York has not always been easy for African Americans. For most of the 20th century cultural segregation was the norm. While Jim Crow laws in Southern states were explicit, here in New York there also were known rules of discriminatory racial separation in accommodations that could make finding a cool place on hot summer days challenging.[14]

If the Adirondack wilderness is to continue to be the New York treasure it has been since the 1880s and '90s, then our park must be welcoming to all New Yorkers. All of us began as neophytes, and if we love wilderness and want it protected, we need to share it. In 1967 Dolly Robinson spoke of how African Americans of her day did not know much about the Forest Preserve. But she wanted to protect it for all New Yorkers. The future of the Forest Preserve depends on the efforts of those who cherish wilderness to see that Dolly Robinson's hopes that the Forest Preserve is for everyone are realized.

That we have a way to go was painfully illustrated in 2016 when Aaron Mair, a Black man and, coincidentally, one of the nation's leading environmentalists, was racially insulted while canoeing on the Schroon River. A group of rafters also on the river that day encountered Mair, invoked a familiar racist slur, and made it clear that they considered Mair an unwelcome intrusion into a paradise they considered their exclusive domain.[15] In 2024, North Country Public Radio ran a piece on right-wing politics in and around the Adirondacks and found, among other things, Confederate flags prominently displayed in many places in the Park. That flag, of course, does not reflect the views of all Adirondackers. When a monster truck displaying the flag across its hood participated in the parade for the 2024 Saranac Lake Winter Carnival, it drew boos and condemnation—along with cheers.[16] Dolly Robinson's vision for an inclusive and diverse enjoyment of our wilderness remains just out of reach.

28

The Constitution, 2038

In 2037, give or take a year or two, New York voters will be asked, as they are roughly every twenty years, whether they want to call a constitutional convention to reform the state's chief governance document. It will have been seventy years since a convention sat to write or revise our constitution, and it's impossible to predict how voters will lean. Such a convention is expensive, and there is always the chance that a new constitution will be no improvement over the last. But New York has never gone this long without a serious attempt to modernize its constitution. If voters approve, the usual pressing issues will present themselves to the delegates, whoever they are: the need to rationalize New York's baroque judiciary, how to fairly draw legislative and Congressional districts, the never-ending struggle between New York City and the rest of the state over power and the allocation of resources, and the efficient construction of a twenty-first-century welfare apparatus, among many others. But Article 14 will also draw significant attention. If the voters in November of 2037 give the nod to a convention, it will gather in Albany in the summer of 2038, with a statewide vote on a new constitution in November.

How will Article 14 be interpreted for a forest where a warming climate is driving changes of species composition, where boreal species living and reproducing for millennia may be unable to survive in a warmer regime, where dead snags are surrounded by the saplings of more southerly species moving north? As more and more winter precipitation falls as rain, will global warming eventually mean that our febrile debates and bitterly contested litigation about snowmobile trails will be irrelevant? Will enthusiasts for all-terrain vehicles, which have

182

a far worse environmental impact than snowmobiles do and are currently banned in the Forest Preserve, lobby for access to snowmobile
trails? Will skiing at Whiteface and Gore remain viable? Will there
be opportunities for ice climbing?[1] Will the delegates, including those
sympathetic to the idea of wilderness as it has evolved in New York
ever since the establishment of the Forest Preserve in 1885, conclude
that the time has come for rethinking the whole concept? Adirondack wilderness has been protected and cherished for many reasons,
and nearly all of its defenders have been grateful that those marks of
human impact they could see around them were at least slowly disappearing. The encroaching forest erasing logging roads, the recovery of
burned acreage, and the reforesting of farmland all meant the Adirondack land was healing itself and reverting to an ever more wild state.
But how do we defend a wilderness doctrine when the signs of anthropogenic impact are moving in the opposite direction and becoming
more pronounced every year?[2]

At every convention since 1894—1915, 1938, and 1967—delegates
have introduced proposals to open up the Forest Preserve to logging
and intense recreational development, to turn the Forest Preserve
into a multiple-use resource managed and exploited as the National
Forests are. These have always been voted down. Will delegates in
2038 do the same? Equally important, now that it is beyond serious
argument that the earth's atmosphere is warming and that weather
patterns have begun to change and will continue to do so far into
the distant future, will delegates consider the implications of climate
change for New York's always fragile and beleaguered wilderness? Climate change is a fact and will put unprecedented demands on Forest
Preserve Management.

The causes of climate change are now well known: the burning of
fossil fuels, the energy-intense and methane-producing nature of industrial agriculture, and the clearing of vast expanses of forest, to name
the most important. When we extract and then refine and burn fossil
fuels—coal, oil, natural gas—we release carbon dioxide, methane, and
other gasses that in the upper atmosphere act like a greenhouse and
trap heat that otherwise would escape into space. When we clearcut

forests, as we have massively done throughout the world, especially and more recently in Amazonia, Africa, Indonesia, New Guinea, and pretty much wherever those forests persist, we make the greenhouse problem worse because those billions of trees could have captured some of the carbon through photosynthesis and thus diminished the warming effect. The capacity of living trees to absorb and then hold some of the atmospheric carbon is important to how we ponder the future of the Forest Preserve in our corner of a warming planet.

Understanding the details of all this, especially for those of us not trained as scientists, is an ongoing and imprecise process. Scientists who study atmospheric chemistry, meteorology, paleoclimatology, wildlife management, and silviculture, among others, still debate, often furiously, some finer points around the edges of the discussion, and the effects of all this vary and will vary from region to region. But the consensus is painfully clear: our planet is warming, and we are in large part responsible. The literature accessible to nonscientists has grown immensely in the last few decades, and it's impossible to read and digest it all. Some of the writers who have clarified my thinking include (to start with two Adirondack authors) Bill McKibben and Curt Stager, along with Elizabeth Kolbert and Michael Mann.[3]

These authors and their many comrades, faced with an environmental catastrophe of unimaginable dimensions, have, perhaps inevitably, adopted an organizational scheme and tone reminiscent of George Perkins Marsh, whose *Man and Nature*, published over a century and a half ago, was so important to the establishment of our Forest Preserve. But the looming apocalypse of climate change with all its attendant calamities of rising temperatures, super storms here and drought there, acidification of the oceans, uncontrollable fires, collapse of the world's fisheries, rising sea levels, and dislocation and likely extinction of many plants and animals, is a predicament far more difficult to address than anything foretold by Marsh. Marsh believed that a well-researched, comprehensive warning about the threat to civilization posed by denuded mountains was the essential start to promoting self-preservation and real change in human behavior. In the case of New York and its Forest Preserve, he appears to have been right: people in authority read the warnings, listened to their constituents, and began adopting responsive policies. But the recent record on climate is not promising: the dire warnings are ubiquitous, thorough, and compelling, but the global reaction has been anemic.

What does the advent of dramatic climate change mean for the wilderness of the Forest Preserve, which so many of us continue to value? I've noted how easy it is for me and others to minimize or even ignore the evidence of human impacts on our wilderness: ancient tote roads in the Sewards, for example. Will we be able to do this when we see increasing numbers of dead trees and the disappearance of species unable to adapt to changing climate? No one knows for certain how these changes in the wilderness will manifest or when and to what extent they will occur. Will they be so gradual that we don't notice? Or will they be sudden and dramatic?

In any case, they have already begun. In early 2024, the New York State Energy Research and Development Authority (commonly known as NYSERDA) released a comprehensive study of how climate change is occurring and how we should prepare for even more dramatic losses: "New York State Climate Impacts Assessment: Understanding and Preparing for Our Changing Climate." This report found what we would expect. By the 2050s, less than three decades from now, Adirondack temperatures will be around 4.6 to 6.6 degrees higher than they were during the period from 1980 to 2010; thirty years after that, they will be even higher. Winters will be warmer. By the 2050s, Lake Placid, the Adirondack town for which we have the best long-term weather data, will have half as many days with temperatures below zero as it did only a few decades earlier. Lakes will freeze later and open up earlier. When more precipitation falls as rain rather than snow and thus runs off faster, the threat of flooding will increase.[4]

Wildlife and forests will be affected. For example, the three-toed woodpecker, a smallish bird that since anyone began keeping records has been sighted here and there in our boreal forests, is now either gone for good or so rare that no one has seen it recently. Others are in decline: black-backed woodpeckers, boreal chickadees, Lincoln's sparrows, gray jays, and yellow-bellied flycatchers, among others. Some

of these species are still around, but they are moving to higher elevations.[5] Throughout New York hemlocks, one of our most common
species, are under attack from the invasive wooly adelgid; warmer
weather means this threat will be even greater.[6]

No one has done more to bring the reality of the climate crisis
home to the Adirondacks than Jerry Jenkins. With his enviable talent
for taking complicated science and making it intelligible to the un-
trained reader, he has deftly and economically described the problem.
If global carbon emissions continue to follow current trend lines, the
Adirondack climate a century or so from now will be roughly similar
to that of Georgia today. If we apply our best efforts to cutting carbon
emissions and adopting non-fossil energy sources, the Adirondacks
will have a climate similar to that of today's West Virginia.[7] As Jenkins takes great pains to show, exactly what this means for our forests
is impossible to predict with any certainty. What is glaringly clear,
however, is that our forests and the creatures that live in them will
be massively stressed. Some of our signature tree species—e.g., red
spruce, tamarack, and sugar maple—are at or near the edge of their
temperature range. Will they disappear, diminish, adapt? Much of the
forests of southern Appalachia are dominated by oaks and hickories.
Will these species move north and replace the maples, spruces, and
birches so familiar to us here in the North? For many centuries, the
High Peaks have been covered with snow for several months of every
year, even when the villages around them endured winter thaws and
lost their snow cover. What will less snow and warmer air mean to the
alpine ecosystems on the iconic one hundred or so acres of the treeless
summits? Will those summits remain treeless?

And then there is fire. The United Nations predicts a 50% worldwide increase in wildfires by 2100.[8] Because of routinely irresponsible
logging practices, the Adirondacks endured massive fires in the early
twentieth century, but since then we have largely been spared the devastating fires that are now a guaranteed annual horror in much of the
American West. With the introduction of sensible logging and the
dedication of state and private resources to preventing fires and quickly
suppressing those that started, the devastation of the early 1900s has

not recurred—despite a period of drought in the 1960s that saw local-
ized fire, state-mandated closing of parts of the Forest Preserve, and
widespread anxiety about what might be in store. Given more than
a century of stable precipitation, mostly conservative harvesting, and
vigilance, most of us have become complacent. The Adirondacks have
become known as the "asbestos forest," and we have come to think that
catastrophic fire is just not an Adirondack issue. But as now-retired
New York Forest Ranger Louis Curth has succinctly pointed out, this
can lead to complacency and an inability to grapple with changing
circumstances. In addition to rising temperatures, the warming global
climate will mean radical and unpredictable changes in precipitation
patterns.[9]

No expert can say what the precise, on-the-ground implications
of climate change and fire are for our constitutionally protected wilderness, but we can expect unconstitutional proposals to deal with an
approaching crisis. Consider what happened after the great blowdown
of 1950, an event that led to sudden changes in Adirondack forests,
mainly hundreds of thousands of acres with downed trees. Soon after
the storm, the New York attorney general issued an opinion that the
state could contract for salvage operations in the Forest Preserve,
something ostensibly forbidden by Article 14. He was responding to
a widely expressed fear that all those downed trees would in a short
time present a massive threat of fire, both in the Forest Preserve and
on private land. In other words, he unilaterally abrogated Article 14.
Environmentalists objected, but no one had the inclination, energy, or
resources to take the matter to court.[10]

It seems possible, even likely, that a warming climate in the Adirondacks will similarly lead to dead trees and an increased threat of
fire. Will those who cherish Article 14 and what it has meant for wilderness in the Forest Preserve keep quiet, give up on even the possibility of wilderness and the natural recovery of stressed ecosystems,
and accept a putative need for new ways of thinking? What we do with
Article 14 in the context of a rapidly warming climate is a matter crucially important in many ways, and we need to start thinking about it
now. Given how well Article 14 has served us up until now, I cautiously

argue for leaving it alone. It's an all-too-certain sign of anthropocentric hubris to assume that a problem that we have created through our manipulation of natural systems, intended or not, can somehow be ameliorated through some further manipulation of that same system.

Whatever the visible impact of climate change on the Forest Preserve, it's important to note the importance of an intact forest. Every year, most trees absorb carbon dioxide from the atmosphere. This is called sequestration. Once that captured carbon is in the tree, it is stored and thus does not contribute to the greenhouse effect. Scientists who study trees and all that they do disagree on exactly how much carbon is withdrawn from the atmosphere, which species and ages do the best job, and how much is stored and for how long, but there is no doubt that the Adirondack Forest Preserve is doing important work and that preventing further loss of intact forests—in the Adirondacks, throughout the state, and around the world—is essential.[11] No discussion of Article 14 should proceed without serious consideration of how it promotes the sequestration and storage of carbon, or of how such a forest can serve as a refuge for species pushed north by warming and fragmented habitat.[12]

In December of 2022, Governor Kathy Hochul signed legislation committing New York to conserving thirty percent of its land and water surface by the year 2030. Known as "30x30," such an achievement would protect biodiversity, wildlife habitats, and clean water and would significantly contribute to our state's efforts to limit climate change. At the time, roughly twenty percent of New York's surface was protected in some fashion: municipal, county, and state parks; state forests; conservation easements; and, most important, the 2.9 million acres of the state Forest Preserve, 2.6 million acres of which are in the Adirondacks (with the remaining 300,000 acres in the Catskills). The Forest Preserve is the most secure of all these. The goal of reaching thirty percent within the next few years is ambitious and perhaps out of reach but it's essential, and the key to success is more protected land in the Adirondacks, especially through an aggressive campaign to enlarge the Forest Preserve.[13]

Adding significantly to the Forest Preserve, of course, will run into the usual opposition—from climate skeptics to the naysayers who routinely claim that the state "can't take care of what it already has" and argue that state-owned wilderness exists only to satisfy the demands of elitist and powerful environmental lobbies. In November of 2023, when I wrote a piece for the *Adirondack Explorer* about the 30x30 plan and its implications for the Forest Preserve, online comments both approved and objected. "The state does not manage these lands well at all," wrote one objector, "and I don't trust them or their land classifications. It appears that more land acquisitions are being pushed by the more radical preservationist organizations that want to force local residents out." Another commenter argued that notwithstanding "elitist groups that don't believe in multiple use," the state domain should be logged. But another took up the challenge and insisted that "the only argument that survives hard analysis for protecting or expanding State land is . . . Wilderness Solely for Wilderness's Sake." Like everything connected with how to manage and even enlarge the Forest Preserve, the argument gets hot.[14]

As our world plunges into the unpredictable and probably chaotic climate future, one thing is certain: forests capture and hold carbon. They remove carbon from the atmosphere every summer, and they sequester much of it thereafter, both in the trees themselves and in the soil. The capacity of forests to grab and hold carbon is a fundamental element of all international climate planning. New York's 30x30 goals, as well as the goals spelled out in the 2019 Climate Leadership and Community Protection Act, depend absolutely on protecting the forests we now have and adding to our total of forested acres.[15] The best way to make sure forests are doing their job is to add them to the public domain and grant them the protection of Article 14. Carbon sequestration thus becomes the twenty-first-century version of watershed protection. In the 1880s and 1890s fears about the viability of the Erie Canal led to a constitutional provision protecting wilderness in the Adirondacks. Today, we ponder the threat of climate apocalypse and see a different utilitarian reason to maintain that provision. In both cases the utilitarian argument bolsters the argument for defending wilderness for its spiritual value.

Our Forest Preserve has been a spiritual and recreational treasure for nearly a century and a half. It has protected our rivers and has been home to myriad forms of wildlife. It has soothed the souls of countless hikers and paddlers. These contributions to New York welfare will continue and expand, while a growing Forest Preserve, with the protections of Article 14 reasserted, plays a vital role in the worldwide struggle to mitigate climate catastrophe.

Notes

Bibliography

Index

Notes

1. Topophilia

1. Tuan, *Topophilia*. The advisor who recommended this book was Professor Rod French, who helped me incalculably when I was trying to figure out what I wanted to do with my dissertation; with this recommendation he showed that he understood keenly the intense connection between me and my topic.

2. Protect the Adirondacks! Inc. v. New York State Dept. of Evntl. Conservation and Adirondack Park Agency, 37 NY3d 73 (2021), https://www.nycourts.gov/reporter /3dseries/2021/2021_02734.htm.

3. Peter Bauer, executive director of Protect, has written a complete history of this important litigation. By the time this book reaches print, I trust it will have been published.

4. Something that confuses anyone from outside New York, along with most people who live in the state, is this: in New York, the lowest court for such litigation, the entry point, is called the Supreme Court. The next level is the Appellate Division, divided geographically into four departments, one for each Appellate District. The final court, the one analogous to the United States Supreme Court, is the New York Court of Appeals.

5. hooks, *Belonging*, 1, 34.

6. Seidule, *Robert E. Lee and Me*. Seidule's father, James M. Seidule, was my high-school American history teacher.

7. As I trace the routes of these overlapping journeys, it may seem that I am evasive or cryptic when it comes to some of the details of my personal life. Those details are important to me, of course, and to the other people involved, mostly unnamed here. But the facts of my first marriage, 1971, and its dissolution, 1996, of my two children (born in 1980 and 1982), and of my second marriage, 2012, add nothing to what I'm trying to say about the Adirondacks. We all have our memories, with both joy and sadness, but none of that is relevant here.

8. Nussbaum, "A Peopled Wilderness," 21.

9. For a thoughtful essay focusing on the Adirondacks, acknowledging all the ambiguities and pitfalls involved with our continuing to use the term and to protect wilderness as such, see Ouderkirk, "On Wilderness and People," 435–60.

10. Colvin, *Seventh Annual Report*, 67.

2. Contact

1. It was only on the occasion of preparing our fiftieth-reunion book that I learned that at least one of my classmates had been sent to EHS by parents eager to keep him out of recently integrated public schools in South Carolina. I'm pleased to say that EHS integrated soon after I graduated—better late than never. See Morris, "Forcing Progress."

2. Murray, *Adventures in the Wilderness*, 52.

3. Bill Verner

1. On maps and in guidebooks, this spot is identified as Plumley Point. At AWC, it was always called Plumley's.

2. The Northville-Lake Placid Trail is mostly north-south, but from Plumley's to Duck Hole, it's more or less east-west.

3. The others, so far as I can recall now, would have been Jerry Swinney, George Bowditch, and Harold Hochschild at the Museum, along with Paul Jamieson, Grace Hudowalski, Jim Goodwin, George Marshall, Mary MacKenzie, and Warder Cadbury. The next few decades would see a rapidly growing number of people interested in Adirondack history, in which club I counted myself.

4. Edmondson, *A Wild Idea*, 162–64.

5. "Report of the Citizens' Advisory Task Force." I own a copy of this document, which was distributed to members of the legislature, the Park Agency, and other stakeholders. McMartin, *Perspectives on the Adirondacks*, 52–53.

6. Nash, *Wilderness*. This was the first of several editions of this book. I inscribed the date I bought it (August 1968) inside the front cover, where a sticker from the Adirondack Museum confirms the place of purchase.

4. Couchsachraga

1. On the rise of camping in the twentieth century, see White, *Under the Stars*; Chamberlin, *On the Trail*; and Young, *Heading Out*.

2. For a brilliant and often quirky assessment of how American reservations about modern civilization have been consistently coopted and colonized by an insidiously aggressive consumer culture, see Lears, *No Place of Grace*.

3. The story about how the Forty-Six High Peaks came to be the focus of a major hiking club, how they were originally thought, in the 1920s, to be the only Adirondack peaks over 4,000 feet high, and how subsequent surveying and mapping by

the US Geological Survey revealed that the initial list was based on errors inscribed on the early twentieth-century maps but was maintained for the sake of tradition is summarized in my Introduction to the Adirondack Mountain Club reprint of Carson's *Peaks and People of the Adirondacks*, xxvii–xxxiii. For more detail, see the three books published by the Forty-Sixers: *The Adirondack High Peaks*; *Of the Summits, Of the Forests*; and *Heaven Up-h'sted-ness*. The routes up the "trailless" peaks were briefly described in the only available guidebook of that era, the Adirondack Mountain Club's *Guide to Adirondack Trails* from 1962; the routes for Cooch and Panther are pp. D-5 and D-6.

4. Pownall, *A Topographical Description*, 51.

5. Marshall, *The High Peaks of the Adirondacks*, 38. This pamphlet was the germ around which Russell Carson organized his *Peaks and People of the Adirondacks*.

6. See Mike Lynch and James Odato, "A Booming Decade of 46er Finishers," *Adirondack Explorer* (May/June 2024): 28, 31.

7. I didn't report in until a few years later, so my number, 772, is a few notches higher than it would have been if I had followed the conventional path to Forty-Sixer-dom, reporting before the end of 1969.

8. The Forty-Sixers, *Heaven Up-h'sted-ness*, 289–92.

5. Adirondack Museum

1. Terrie, *Wildlife and Wilderness*.

2. Hoffman, *Wild Scenes in the Forest and Prairie*; Headley, *The Adirondack*; Hammond, *Hills, Lakes and Forest Streams*; Street, *The Indian Pass* and *Woods and Waters*; Thorpe, "A Visit to John Brown's Tract"; Murray, *Adventures in the Wilderness*. Bill Verner called the years from 1840 to the American Civil War the "Golden Age" of Adirondack sporting literature. See my "Classic Adirondack Literature" in *Seeing the Forest*, 99–104.

3. Colvin's account of his "discovery" of Lake Tear in 1872 was published for the first time in Carson, *Peaks and People*, 138–44.

6. Roderick Nash

1. Bradford, *Of Plymouth Plantation*, 69–70.

2. Nash, *Wilderness*, 24.

3. See, among many others, Cronon, *Changes in the Land*; Jennings, *The Invasion of America*, 30. I'm grateful to Pete Nelson for sharing Jennings's insight with me.

4. Slotkin, *Regeneration Through Violence*, 3–93.

5. Nash, *Wilderness*, 53.

6. Nash, *Wilderness*, 54–55.

7. Bartram, *Travels*, 115–24.

8. Quoted in Nash, *Wilderness*, 23.

9. Headley, *The Adirondack*, 167–68.

10. Headley, *The Adirondack*, 1.

11. See my Introduction to the 1982 reprint of Headley, *The Adirondack*, 5–10.

12. Nash, *Wilderness*, 44–66.

13. See, among many others, Novak, *American Landscape and Painting*, which takes a decidedly Nashian approach to the study of American landscape painting.

14. Nash, *Wilderness*, 67–83; Sears, *Sacred Places*, 78–80, 144–45.

15. Nash, *Wilderness*, 122–40. See also Fox, *John Muir and His Legacy* and Cohen, *The Pathless Way.*

16. Nash, *Wilderness*, 161–81.

17. See Runte, *National Parks*, 1–64.

18. See Worster, *Nature's Economy*, 258–90.

19. Nash, *Wilderness*, 182–99.

20. Runte, *National Parks*, 128–36.

21. Nash, *Wilderness*, 183–89.

22. Meine, *Aldo Leopold*, 243–48.

23. Nash, *Wilderness*, 201–25. See also Harvey, *Wilderness Forever.*

24. Nash, "Wilderness and the American Mind: Fifth Edition," Yale Univ. Press (last accessed October 31, 2024), https://yalebooks.yale.edu/book/9780300190380 /wilderness-and-the-american-mind/.

25. For example, see Ryan, *This Land Was Saved*, a faithful rehearsal of the Nash paradigm.

7. Robert Marshall

1. In my files I have a letter to me from George Marshall (June 30, 1979) confirming the influence of the Colvin reports on him and his brother. In 2000 I briefly entertained the idea of writing a biography of Robert Marshall and spent a week reading his letters and other papers housed at the Bancroft Library of the University of California at Berkeley. Later, after discovering how well Paul Sutter had told the story of Robert Marshall and his achievements in *Driven Wild* and concluding that he had done it all much better than I ever would, I abandoned the project.

2. Marshall, *The High Peaks of the Adirondacks*. This was also the first publication of the months-old ADK.

3. Marshall, "The Problem of the Wilderness," 141–48. See Glover, *A Wilderness Original* and Sutter, *Driven Wild*, 194–238. The other founders of the Wilderness Society were Aldo Leopold and Benton MacKaye. On the history and activism of TWS, see Sutter, *Driven Wild.*

4. Marshall, *Alaska Wilderness*. This book was published after Marshall's death; George Marshall, who catalogued his brother's voluminous correspondence and

journals, assembled Bob's Alaska exploration narratives for publication in book form. Bob's other Alaska book, published while he was still very much alive, was *Arctic Village*, a sociological assessment of a population of prospectors and Innuit in Wiseman, Alaska, where he conducted silviculture research for fifteen months, 1929–30.

5. Marshall, *Alaska Wilderness*, 40.

6. Glover, *A Wilderness Original*, 17.

7. Marshall, *Arctic Wilderness*, 1.

8. See Johnson, *Edward Stratemeyer and the Stratemeyer Syndicate*, 1–17. My student (now professor) Ilana Nash steered me to the useful sources on Stratemeyer and his extensive children's literature domain.

9. *Newark Sunday News*, ca. 1904, quoted in Johnson, *Edward Stratemeyer and the Stratemeyer Syndicate*, 5–6.

10. Bonehill, *Pioneer Boys*, 231.

11. Bonehill, *Pioneer Boys*: Indians are "marauding savages" on p. 3, "rascals" on p. 32, "thieving" on p. 42, "treacherous" on p. 47, "dirty, foul-smelling" on p. 150.

12. Marshall, *Arctic Wilderness*, 2–3.

13. Marshall, *Arctic Wilderness*, 3.

14. The novel is in box 15 of the Marshall Papers. In the same box is a letter (June 4, 1931) from Marshall to Lenore Marshall, his brother Jim's wife, asking her to find a typist and submit the manuscript to "the publishing house of Jonathan Cape and Harrison Smith." He added, "When they reject it would you be kind enough to keep it in safe storage for me until I return." I read the manuscript and concluded that it was dreadful, quite unpublishable, and that any editor would have quickly rejected it. The fact that Marshall was still rereading *Pioneer Boys* when he was twenty-one shows how his strengths as a writer lay in his tireless and enthusiastic conservation advocacy, not fiction.

15. The chronology here doesn't really work. Colvin first explored the Bog River in 1871, twelve years after John Brown's burial. See Rosevear, *Colvin in the Adirondacks*, 3.

16. Marshall, *Arctic Wilderness*, 32.

17. Marshall, *Arctic Wilderness*, 31, 40.

18. In "The Problem of the Wilderness," he asserts that "when Columbus effected his immortal debarkation, he touched upon a wilderness which embraced virtually a hemisphere." He acknowledges the Indian presence but seems unaware of their modifications of the land. Was Marshall not aware that Native Americans skillfully and routinely manipulated the various environments in which they lived? He does acknowledge "their trails and temporary shelters."

19. Kaye, "Conservation Controversy." My photocopy of this article does not have a date.

20. Unless otherwise indicated, all quotations in this section are from Marshall's "The Problem of the Wilderness."

21. See Sutter, *Driven Wild*, 195, 210–11, 237–38.

22. See, among others, Mann, *1491*.

23. In his essay on wilderness, which I discuss in chapter 15, William Cronon carefully analyzes how such views as Marshall's removed Indigenous people from history.

24. Longstreth, *The Adirondacks*, 7–8.

25. Donaldson, *History*, vol. 1, 21. For a compact and excellent summary of the seasonality problem, see Reynolds, "Early Nomads or Occupants of the Adirondacks?," 32–34. On Champlain's 1609 encounter, see Morison, *Samuel de Champlain*, 109–11.

26. Otis, *Rural Indigenousness*.

27. The archaeological evidence for a Native presence in the Adirondacks is strong and growing stronger. See Stager, "Hidden Heritage."

28. Cronon, *Changes in the Land*, 58–59; Hämäläinen, *Indigenous Continent*, 197–98.

29. See Sutter, *Driven Wild*, 236–38.

8. Forever Kept as Wild Forest Lands

1. For a start at counting the amendments to this provision, see VanValkenburgh, *The Adirondack Forest Preserve*.

2. Marsh, *Man and Nature*. Marsh's conclusions were largely based on his observations around the Mediterranean, where he was posted as a diplomat in the years leading up to the American Civil War. He believed—and amassed copious evidence to demonstrate—that the decline and collapse of ancient civilizations, especially the Roman Empire, could be explained by profligate destruction of forests on mountain slopes. Recent work in the field of climate history confirms that irresponsible forest clearing was a factor but that it must be assessed in the context of natural cycles of climate change. The "Roman Climate Optimum," as the period between the late Republic and the reign of Marcus Aurelius has come to be known, and which was a period of widespread economic and political stability in the Roman world, ended roughly near the beginning of the third century CE. After that, precipitation declined, temperatures cooled, harvests became less predictable, and the quality of life deteriorated in much of the Empire. See Harper, *The Fate of Rome*.

3. Laws of New York, 1872, chapter 848; New York State Commissioners of State Parks, *First Annual Report* (Senate Document 102, 1873).

4. Laws of the State of New York, United States: n.p., 1883, chapter 13, https://www.google.com/books/edition/Laws_of_the_State_of_New_York/3p04AAAA IAAJ.

5. That these lands, most of which had for centuries been occupied or used by various Iroquoian—Haudenosaunee—and Algonquian communities, were considered by White authorities to be legally owned by the state—rather than stolen—and thus a commodity that could be bought and sold is another matter. See chapter 7 on Robert Marshall.

6. An excellent summary of state policy from the Revolution up to 1883, when the State adopted a radically changed policy, from "sell it all, as quickly as possible" to "keep what we have," can be found in Lincoln, *The Constitutional History of New York*, vol. 3, 392–417. All five volumes of this monumental work are online courtesy of the New York State Library, https://nysl.ptfs.com/#!/s?a=c&q=*&type=16&criteria=field11%3D1337955&b=0.

7. Chapter 551, Laws of 1884, https://hdl.handle.net/2027/ucl.b2998351.

8. See Edmondson, *A Wild Idea*.

9. Assembly Document 36, 1885. The quotations in the remainder of this chapter, unless otherwise identified, are from this document.

10. See the section on Colvin in Adirondack Mountain Club, *Adirondack Bibliography*, 47.

11. Pyne, *Fire in America*, 199–201.

9. The Forest Preserve

1. This is obviously a key piece of legislation, but it is available online only at sites where a subscription is required. Printed volumes of session laws are available at most New York law libraries. Unless indicated otherwise, all quotations in this chapter are from this statute, a complete copy of which I have in my files. Another law of that year, Chapter 448 of the Laws of 1885, pursued the Sargent Commission's recommendation that the State clarify and tighten the statutes under which the State acquired real property at tax auctions and confirmed titles.

2. *New York State Conservationist*, May–June 1985.

3. All New York constitutions can be found online. The one approved in 1846 is here: https://images.procon.org/wp-content/uploads/sites/48/1846_ny_constitution.pdf.

10. The Adirondack Park

1. Nash, *Wilderness*, 120.

2. Chapter 707, Laws of the State of New York, 1892.

3. For a thorough discussion of how the word "Park" was chosen in 1892 and what it meant with respect to the already-established Forest Preserve, see Lincoln, *The Constitutional History of New York*, vol. 3, 422–27.

4. Chapter 332, Laws of the State of New York, 1893, Section 103.

5. Lincoln, *The Constitutional History of New York*, vol. 3, 428–29.

11. The Constitution, 1894

1. The provision of New York law that voters be asked every twenty years whether they want to call a constitutional convention was written into the constitution of 1846 and has been in effect ever since. See Galie, *Ordered Liberty*, 109–10. On the 1894 convention, see 159–87.

2. Donaldson, *History*, vol. 2, 173.

3. Donaldson, *History*, vol. 2, 188.

4. The pamphlet was printed without any indication of place of publication, but it is dated June 1894 on the cover; on the first page the date is June 13, 1894. Available online at https://hdl.handle.net/2027/mdp.35112105253282. I found a copy for sale on AbeBooks as a reprint issued by Pranava Books of India.

5. Section 3 is concerned with establishing a salaried state agent, a "Superintendent of Forest Preserves" to manage the Forest Preserve and see that all laws concerning it were properly executed. This person would be empowered to hire others to help with his duties (it was assumed at the time that only men would have these jobs). Section 4 stipulates that the Superintendent would have the power to lease state lands in parcels of no more than five acres for clubs and camps. This is a subject that would vex and confuse Adirondack enthusiasts until after the adoption of the State Land Master Plan in 1972.

6. NYBTT, "The Proposed Amendment," 18. Marsh is cited on 7 and quoted on 17.

7. Laws of the State of New York, 1893, Chapter 332; NYBTT, "The Proposed Amendment," 24.

8. NYBTT, "The Proposed Amendment," 20–21.

9. NYBTT, "The Proposed Amendment," 24–25.

10. *Revised Record of the Constitutional Convention of the State of New York, May 8, 1894, to September 29, 1894*, 5 vols. (Argus Co.: 1900). Vol. 1 is 1,191 pages; vol. 2 is 1,202 pages; vol. 3, 1,242 pages; vol. 4, 1,280 pages; vol. 5, which is related documents, is 1,117 pages. Cited hereafter as *Rev Rec 1894*, [vol.], [page]. The citation for the provision proposed by McClure is *Rev Rec 1894*, 4, 124. See also Lincoln, *The Constitutional History of New York*, vol. 3, 430–35. All five volumes of the *Revised Record* are available online in Google Books. Vol. 4, the key volume for our purposes can be found at https://www.google.com/books/edition/Revised_Record_of_the_Constitutional_Con/SG00AQAAMAAJ?hl=en&gbpv=1&printsec=frontcover. Another important source, unjustifiably neglected, on the history of the Forest Preserve and the state constitution is "Article XIV" by Ralph Semerad. This is one of the Technical Reports submitted to the Temporary Study Commission on the Future of the Adirondacks in 1970. These reports were published by the Adirondack Museum in 1971 in a volume titled *The Future of the Adirondacks: The Reports of the Temporary Study Commission on the Future of the Adirondacks*.

11. Donaldson, *History*, vol. 2, 190. Donaldson also refers to him as "Colonel McClure."

12. *New York Times*, May 1, 1912. It's possible of course that McClure had visited the Adirondacks. Peter Bauer has told me that he heard of this when he worked at *Adirondack Life*, but I have been unable to track down any evidence.

13. Many of the New York City Social Registers can now be found online, courtesy of Google Books. See, for example, the edition of November 1895, at https://www.google.com/books/edition/Social_Register_New_York/q3wBAAAAYAAJ?hl=en&gbpv=1.

14. *Rev Rec 1894*, 4, 124.

15. *Rev Rec 1894*, 4, 129–30.

16. *Rev Rec 1894*, 4, 132, 134.

17. *Rev Rec 1894*, 4, 139.

18. The word "ecosystem" was first used by ecologist George Tansley in 1935. See Worster, *Nature's Economy*, 378.

19. *Rev Rec 1894*, 4, 141–43.

20. *Rev Rec 1894*, 4, 144.

21. Jacoby, *Crimes against Nature*, 52–54, 56–57, 64, 66.

22. *Rev Rec 1894*, 4, 147–49. Delegate Mereness represented Lewis County. Two other members of the committee that initiated the Forest Preserve provision (and was chaired by David McClure) were from northern New York: John McIntyre of St. Lawrence County and Chester McLaughlin of Essex County. See Donaldson, *History*, vol. 2, 191.

23. *Rev Rec 1894*, 4, 149.

24. *Rev Rec 1894*, 4, 150–52.

25. *Rev Rec 1894*, 4, 154–56.

26. *Rev Rec 1894*, 4, 157–58.

27. *Rev Rec 1894*, 158–61; Donaldson, *History*, vol. 2, 193.

12. Timber

1. Brief for Appellants-Respondents, June 10, 2020, at the New York Court of Appeals, in APL-2012-00196, p. 43. I have a copy of this brief in my files. It can also be found on Protect's website: https://www.protectadks.org/wp-content/uploads/2020/09/ProtecttheAdirondacksvNYSDeptofEnvtlCons-app-NYSDeptofEnvtlCons-brf.pdf.

2. *Webster's International Dictionary* (1892).

3. *Historical Dictionary* (1991), 179.

4. Headley, *The Adirondack*, 76.

5. Colvin, *Report of the Superintendent*, 90–91. These "chutes" became the tote roads whose slowly disappearing vestiges I followed nearly a century later.

6. Lincoln, *The Constitutional History of New York*, vol. 3, 433.

7. Lincoln, *The Constitutional History of New York*, vol. 3, 440.

13. *Forever Wild*

1. The literature on the origins of American Studies in English Departments is vast. For a succinct and witty distillation of the tale, see Wise, "'Paradigm Dramas,'" 294–377.

2. Smith, *Virgin Land*. See Kuklick, "Myth and Symbol in American Studies," 435–50.

3. Joanna left GWU, where she did not have a secure job, while I was in my second year and accepted a tenure-track position at Skidmore College. One day during the time when I was more or less at sea about a dissertation topic and was on the way home from Long Lake, I stopped to see her in Saratoga Springs. She fed me lunch and listened to my confusion: I admired Nash and was obsessed with the Adirondacks, but after Nash, what was there to say? She observed that a legitimate dissertation could be a case study, using the Adirondacks as primary focus and asking whether Nash's paradigm fit. It's blindingly obvious that she was right, but I was slow to get there.

4. Conaway, *America's Library*.

5. The first volume was published by the Adirondack Mountain Club (ADK) in 1958. Both it and the second volume were projects of an ADK committee, but in both cases nearly all of the heavy lifting was done by Dorothy Plum. After Bill Verner, no one has helped me more with my scholarship than has Dorothy Plum, whom I never had the pleasure of meeting; she has long since passed away. I herewith express my eternal gratitude for her prodigious and meticulous labors. Adirondack Mountain Club, *Adirondack Bibliography*, and Adirondack Mountain Club, *Adirondack Bibliography Supplement*. See also Welch, *Adirondack Books, 1962–1992*. For thirty years, there has been no further update.

6. "Urban Man Confronts the Wilderness," 7–20. Barney Mergen died in February 2024.

7. Nash, *Wilderness*, 44–66.

8. The overemphasis on the watershed argument, at the expense of spiritual values, appeared prominently in the first article I drew out of the dissertation, "The Adirondack Forest Preserve: The Irony of Forever Wild," where the very title underplayed the idea of the Forest Preserve as a spiritual asset. Terrie, "The Adirondack Forest Preserve," *New York History*, 261–88.

9. Terrie, "The Adirondack Forest Preserve," *New York History*; Terrie, "Romantic Travelers," *American Studies*, 59–75; Terrie, "Verplanck Colvin, Adirondack Surveyor," *Environmental Review*, 275; Terrie, "The New York Natural History Survey," *Journal of the Early Republic*, 185–206.

10. The 1964 Wilderness Act defines wilderness thus: "A wilderness, in contrast with those areas where man and his works dominate the landscape, is hereby recognized as an area where the earth and its community of life are untrammeled by man, where man himself is a visitor who does not remain. An area of wilderness is further defined to mean in this Act an area of undeveloped Federal land retaining its primeval character and influence, without permanent improvements or human habitation, which is protected and managed so as to preserve its natural conditions and which (1) generally appears to have been affected primarily by the forces of nature, with the imprint of man's work substantially unnoticeable; (2) has outstanding opportunities for solitude or a primitive and unconfined type of recreation; (3) has at least five thousand acres of land or is of sufficient size as to make practicable its preservation and use in an unimpaired condition; and (4) may also contain ecological, geological, or other features of scientific, educational, scenic, or historical value."

The Adirondack State Land Master Plan, approved by Governor Nelson Rockefeller in 1972, reads thus: "A wilderness area, in contrast with those areas where man and his own works dominate the landscape, is an area where the earth and its community of life are untrammeled by man—where man himself is a visitor who does not remain. A wilderness area is further defined to mean an area of state land or water having a primeval character, without significant improvement or permanent human habitation, which is protected and managed so as to preserve, enhance and restore, where necessary, its natural conditions, and which (1) generally appears to have been affected primarily by the forces of nature, with the imprint of man's work substantially unnoticeable; (2) has outstanding opportunities for solitude or a primitive and unconfined type of recreation; (3) has at least ten thousand acres of contiguous land and water or is of sufficient size and character as to make practicable its preservation and use in an unimpaired condition; and (4) may also contain ecological, geological or other features of scientific, educational, scenic or historical value." See https://apa .ny.gov/State_Land/Definitions.htm.

11. Nash, *"Forever Wild"* (Book Review), *American Historical Review*, 478–79.

12. Terrie, *Forever Wild: Environmental Aesthetics*, v, 104.

13. Terrie, *Forever Wild: A Cultural History*.

14. Long Lake

1. Population data supplied by James McMartin Long, statistician for Protect the Adirondacks! He harvested the numbers for 1970 to 2020 from US Census data and shared them with me via personal correspondence.

2. The number fifty-nine comes from a personal communication with Alex Roalsvig, director of Long Lake's Office of Parks and Recreation.

3. Details on this project are in the Long Lake Town Archives, where they were discovered and organized by Abbie Verner.

4. Because of its small size and the fact that much of the school budget is based on taxes paid by the State on the Forest Preserve and by the owners of vast forest lands—like Whitney Park—the Long Lake Central School often has the highest expenditure per student of any district in New York. In 2022–23, for example, the latest year for which I could find figures, with a K–12 population of fifty-nine, the per-student cost at LLCS was $69,880, the highest in the state. See https://www .newyorkupstate.com/schools/2024/07/see-the-25-upstate-ny-school-districts-with -the-highest-per-pupil-spending.html.

5. For more on the demographic challenges confronted by Adirondack school districts, see Sara Foss, "Shrinking Enrollments, Creative Solutions: Adirondack Schools Cope with Declining Populations," *Adirondack Explorer* (January/February 2022): 20–23.

6. Adirondack Park Regional Assessment Project, "*The Adirondack Park*: Seeking Balance."

7. See the decision in *Helms v. Reid*, issued by the Supreme Court of Hamilton County in 1977: https://casetext.com/case/helms-v-reid.

8. See "Whitney Property on Little Tupper Lake Is Up for Sale," *Adirondack Daily Enterprise* (November 9, 2023).

9. I do not know if this is true, nor do I know the names of anyone involved.

10. Long and Bauer, *The Adirondack Park and Rural America*.

15. William Cronon

1. Cronon, "The Trouble with Wilderness," *New York Times Magazine*. Also published in Cronon, *Uncommon Ground*, 69–90. All my references to and quotations from this essay use the version in *Uncommon Ground*, which is largely, though not entirely, the same as that in the *New York Times*. The differences between the two are strictly of style and organization. Substantively, they are the same.

2. For a critique of this gendered assessment of the masculine roots of wilderness appreciation, see Schrepfer, *Nature's Altars*.

3. See, among others, Spence, *Dispossessing the Wilderness*, and Jacoby, *Crimes Against Nature*. Jacoby, in addition to exploring the removal of Native Americans from what became Yellowstone and Grand Canyon National Parks, provides an excellent, though arguably overstated, account of how the imposition of statutory and constitutional protections in the Adirondacks at the end of the nineteenth century focused on the recreational needs of elite, downstate sportsmen and vacationers and ignored the existence and economic needs of the year-round population.

4. Cronon, *Uncommon Ground*, 171–85.

5. *The Nation*, April 8, 1996, and June 7, 1996.

6. For a succinct summary of the wide range of responses to Cronon's instantly controversial article, see Denevan, "The 'Pristine Myth' Revisited," 576–91. The quotation from Luther Standing Bear is on 576.

7. Headley, *Washington and His Generals*, a transparent argument for American expansionism; Headley, *The Great Riots of New York*, where he argues that all riots were the fault of immigrants and that rioters should be shot on sight. Headley was elected New York's secretary of state in 1855 on the Know-Nothing ticket; the Know-Nothings were anti-Catholic, chauvinistic, and xenophobic. See my Introduction to the 1982 Harbor Hill reprint of *The Adirondack*.

8. On Audubon and his racial views, including the fact that he owned and sold slaves, see Nobles, *John James Audubon*, 51, 53–55, 161, 164, 201, 202–3, 212. On Muir, see, for example, Lucy Tompkins, "Sierra Club Says It Must Confront the Racism of John Muir," *New York Times*, July 23, 2020. The story on Muir is complicated; see also Michelle Nihuis, "Don't Cancel John Muir," on the website of *The Atlantic*, April 12, 2021, https://www.theatlantic.com/ideas/archive/2021/04/conservation-movements-complicated-history/618556/.

9. The scholarly and popular assessment of the Cronon essay and the ensuing debates over the cultural construction of wilderness are thoroughly examined in two invaluable collections: Callicott and Nelson, *The Great New Wilderness Debate*, and Nelson and Callicott, *The Wilderness Debate Rages On*.

10. Sutter, *Driven Wild*, 13 and passim. For a vigorous presentation of the idea that protection of the wilderness in the Adirondacks was an overt expression of class privilege exercised at the expense of the lives and welfare of the working-class, year-round residents, see Halper, "'A Rich Man's Paradise,'" 193–267.

16. Wilderness and American Studies

1. Marx, *The Machine in the Garden*; Ward, *Andrew Jackson*.

2. Cronon does not mention Nash in "The Trouble with Wilderness," nor does he explicitly critique the myth-symbol school, but the spirit of Nash and *Wilderness and the American Mind* hovers over nearly every paragraph.

3. Kolodny, *The Lay of the Land* and *The Land Before Her*. *The Lay of the Land* is based on Kolodny's dissertation at Berkeley, where one of her graduate advisors was Henry Nash Smith, whom she gratefully acknowledged (x) as a mentor in *The Lay of the Land*, a book whose title deftly echoes Smith's *Virgin Land* and simultaneously emphasizes Smith's failure to address issues of gender in his analysis.

4. Once Cronon introduced his challenge to the Nash thesis, he quickly inspired a host of others to extend and occasionally rebut his position. See especially the section on "Race, Class, Culture, and Wilderness" in Nelson and Callicott, *The Wilderness Debate Rages On*.

5. It has always struck me as a point of interest that the subtitle of Thoreau's *Walden*—"or, Life in the Woods"—mirrors so precisely the subtitle of Headley's *The Adirondack*—"or, Life in the Woods." Headley's book was published in 1849, *Walden* in 1854. Given what we know about Thoreau's voracious reading habits, it seems reasonable to assume that he had encountered Headley's *Adirondack*.

6. Terrie, *Forever Wild*, 23.

7. Gordon-Reed, *The Hemingses of Monticello*, 466–67. James Madison, who accompanied Jefferson and Hemings on this excursion, paid little attention to the scenery but wrote in his diary that on the east side of Lake George they came upon a farm "owned & inhabited by a free Negro. . . . He possesses a good farm of about 250 Acres which he cultivates with 6 white hirelings for which he is said to have paid about 2 ½ dollrs. per Acre and by his industry & good management turns to good account. He is intelligent; reads and writes & understands accounts, and is dextrous in his affairs." The Madison diary is quoted by Gordon-Reed on 466. As Gordon-Reed notes, it would be wonderful to know what all three of these men thought of the fact that a Black man could be peacefully living out the American dream of the "self-sufficient yeoman farmer with a basic education who was able to become a functioning and productive member of society." For Jefferson and Madison, the existence of this farmer "was a total rebuke to [their] way of life on every level." To Hemings, whose response we cannot know, this farmer must have presented a profound and intriguing contrast to the agricultural regime he was familiar with in Virginia.

8. Quoted in Miller, *This Radical Land*, 67.

9. Otis, *Rural Indigenousness*, 103–13.

17. *Contested Terrain*

1. Tyler and Wilson, "Adirondack Stories: Historical Narratives of Wilderness," 21–49.

2. Miller, "Witness Tree," 92. This dissertation was expanded and revised in book form as *This Radical Land*.

3. The idea of the middle landscape was a much-explored trope in the myth-symbol days. The classic assessment of its cultural significance in the United States, at least in white, elite culture, remains Marx's *The Machine in the Garden*.

18. The Constitution, 1915

1. The referendum for holding a convention was in fact early but was moved up by a Democratic governor and a Democratic-controlled legislature to try to ensure that their party controlled the apportionment machinery leading up the Congressional election of 1916. See Galie, *Ordered Liberty*, 188. The primary document for this discussion of the 1915 convention is the *Revised Record of the Constitutional Convention of the State of New York, April Sixth to September Tenth 1915*, paginated continuously:

vol. 1 is 1–1120, vol. 2 is 1121–2240, vol. 3 is 2241–3360, vol. 4 is 3361–4510, including index. Cited hereafter as *Rev Rec 1915*, with volume number and page; e.g., *Rev Rec 1915*, 2, 1532 means vol. 2, page 1532. The record of the 1915 convention is online at https://catalog.hathitrust.org/Record/001156630.

2. *Rev Rec 1915*, 2, 1327–42.

3. *Rev Rec 1915*, 2, 1329–30.

4. The Conservation Commission was renamed the Conservation Department in 1927, and in 1970 this was rolled into the newly established Department of Environmental Conservation.

5. *Rev Rec 1915*, 2, 1340.

6. *Rev Rec 1915*, 4, 4259.

7. *Rev Rec 1915*, 2, 1340.

8. *Rev Rec 1915*, 2, 1345–46, e.g. The clouded or uncertain titles to property on the shores of Raquette Lake remained a vexing problem for both state land managers and putative owners of camps until 2013, when an amendment to the constitution was passed in an effort to settle the claims of all parties.

9. *Rev Rec 1915*, 2, 1445.

10. Ferris J. Meigs composed a two-volume history of the Santa Clara Lumber Company in 1941. This was not published and exists as a typescript, the only copy of which that I know of is in the library of the Adirondack Experience. How this manuscript came to the library can be found in "Story of Santa Clara Lumber Co, as Told by Late Ferris J. Meigs, Recalls Events Which Influenced Early Tupper History," *Tupper Lake Press and Herald*, December 10, 1970. My thanks to Jenny Ambrose for sharing this clipping.

11. *Rev Rec 1915*, 2, 1447–55.

12. Delegate Mereness, who was on the conservation committee in 1894, recollected David McClure and argued for selling off detached parcels (*Rev Rec 1915*, 2, 1461–63). Delegate Beach argued against Angell, invoking the watershed and canal argument (*Rev Rec 1915*, 2, 1464–69). Delegate Tierney favored Angell and argued that loggers could now be trusted (*Rev Rec 1915*, 2, 1469–71). Delegate Whipple opposed Angell (whom he identified as "the attorney for a lumber company"), invoking the familiar watershed argument and the importance of the Adirondacks as a vacation destination and a retreat for people with tuberculosis. He was opposed to any cutting: "If you cut one tree in the forest you let in a little more air, a little more light and you get a little more evaporation. Now if you cut a thousand trees, and you have multiplied that by one thousand and so on, to the end" (*Rev Rec 1915*, 2, 1471–88). Delegate Dunlap declared that logging could be safe and cited the example of the Adirondack League Club. (On the role of the Adirondack League Club, a huge domain near Old Forge, in promoting responsible logging, see Terrie, "'The Grandest Private Park,'" 73–111.) But he believed the time was not yet right for

logging the Forest Preserve. He favored more campsites and roads (*Rev Rec 1915*, 2 1490–96). Delegate Cobb argued that loggers could not be trusted (*Rev Rec 1915*, 2, 1496–98). Delegate Landreth argued that the constitution was too rigid and was designed to favor the few who already had access; he was opposed to "locking up" the forests (*Rev Rec 1915*, 2, 1503–7). Delegate Meigs (an officer of the same company for which Angell was the attorney, the Santa Clara Lumber Co.) supported Angell's position with a long defense of logging and how the State would eventually need to harvest the resources of the Forest Preserve (*Rev Rec 1915*, 2, 1507–23). Delegate Austin, worried about deforestation and deterioration of the watershed, opposed Angell's proposal: "I would not permit a single living stick of timber to be cut, and I would not permit a dead or downed tree to be taken out or carted away and sold or have anything else done to it. But I would allow these dead and down and living trees to stay where they are; to die if they may, and fall and rot, and, as they have done for countless ages, form a new soil out of which a new forest is ever springing, and form a humus from which streams of crystal water flow" (*Rev Rec 1915*, 2, 1523–29). That's a ringing endorsement of "wilderness" in its modern usage. Delegate Parsons opposed the Angell amendment (*Rev Rec 1915*, 2, 1529–30). Delegate Clinton opposed Angell (*Rev Rec 1915*, 2, 1530–31). Louis Marshall opined, "The adoption of section 7 of article 7 of the constitution which preserved in their wild state the Adirondack and Catskill forests" was "the most important action of the convention of 1894." Likewise, he continued, the most important question in 1915 was also this provision. He argued that the chief concern was watershed. He reminded the delegates that Angell and Meigs were tied to the logging industry (*Rev Rec 1915*, 2, 1531–37).

The debate on the third reading of the new article begins on page 3650 of volume 4. Delegate A. E. Smith, apparently absent when the conservation committee first reported its recommendation, rose to oppose it as "setting the clock of progress, in the matter of the development of our natural resources, back at least 10 years." He was mostly concerned with the development of hydroelectric power and the process for appointing the proposed Commission (*Rev Rec 1915*, 2, 3650–54).

13. *Rev Rec 1915*, 2, 4236.

14. The vote approving the new constitution, with "trees and" placed before "timber" in Article 7, Section 2, is recorded on p. 4332 of vol. 4.

15. Louis Marshall, "Defends Conservation Article in New Constitution," *New York Times*, October 24, 1915. Peter Bauer discovered this article and shared it with me and John Caffry during the preparation for the first round of litigation in *Protect*—but not in time to get it accepted as evidence.

16. On Louis Marshall's distinguished career as a constitutional lawyer and defender of minority rights, see Silver, *Louis Marshall*.

17. "The Budgetary Provisions of the New York Constitution," *The Annals of the American Academy of Political and Social Science* 62 (November 1915): 68, quoted in Galie, *Ordered Liberty*, 200.

19. MacDonald

1. A PDF of this decision can be downloaded here: https://casetext.com/case /association-protection-adirondacks-v-macdonald. Anyone interested in learning more about the steps leading up to this decision and its implications should start with the account in Graham, *The Adirondack Park*, 184–207. My quick summary of the details leading up to the decision at the Court of Appeals is based on Graham. All quotations from the decision are from the online PDF. See also Hopsicker, "Legalizing the 1932 Lake Placid Olympic Bobrun."

2. See Terrie, *Contested Terrain*, 134–40.

3. Why the court said "State Park" in this sentence escapes me; it appears to have been confusing the Forest Preserve with the Park—a confusion that continues to vex many discussions of Adirondack matters. I have talked to members of the New York Assembly who routinely used "Park" and "Forest Preserve" as if they were synonymous.

20. The Constitution, 1938

1. What follows is similar to an article I wrote for the *Adirondack Explorer* in 2017, subsequently republished with only minor changes in my book *Seeing the Forest*. Unless otherwise indicated, all the quotations in this section are from the documents generated by the 1938 convention and housed at the New York State Archives in Albany.

2. See VanValkenburgh, *The Adirondack Forest Preserve*, 123, 144.

3. VanValkenburgh, *The Adirondack Forest Preserve*, 133, 149, 159–60, 163–64.

4. Galie, *Ordered Liberty*, 230–31.

5. Galie, *Ordered Liberty*, 230–31; O'Rourke and Campbell, *Constitution-Making in a Democracy*, 62–70. The *Times* editorial, quoted in O'Rourke and Campbell, ran on October 27, 1936.

6. Sources here and below are boxes L0095 and L0076 at the New York State Archives.

7. *New York Times*, Dec. 11, 1938.

21. The Constitution, 1967

1. On the 1967 convention, see Dullea, *Charter Revision in the Empire State*, and Galie, *Ordered Liberty*, 307–31.

2. Van Valkenburgh, *The Adirondack Forest Preserve*, 168–241.

3. Graham, *The Adirondack Park*, 208–18; Dullea, *Charter Revision in the Empire State*, 246.

4. *Proceedings of the Constitutional Convention of the State of New York, April fourth to September twenty-sixth, 1967*. Twelve volumes, consecutively paginated. Hereafter cited as *Proceedings, 1967*. It can be found online at the New York State Library: https://nysl.ptfs.com/#!/s?a=c&q=*&type=16&criteria=field11%3D19122&b=0. Bergan's proposal begins on p. 521.

5. *Proceedings, 1967*, vol. II, Record, Part I, 525.

6. Graham, *The Adirondack Park*, 212–15.

7. *Proceedings, 1967*, vol. II, Record, Part I, 526.

8. *Proceedings, 1967*, vol. II, Record, Part I, 531–32.

9. *Proceedings, 1967*, vol. II, Record, Part I, 527.

10. Brief biographical sketches of all the delegates, including party affiliation, can be found in *Proceedings, 1967*, vol. II, Record, Part I, 11–106. Dullea, *Charter Revision in the Empire State*, 351–60.

11. *Proceedings, 1967*, vol. II, Record, Part I, 542; Dullea, *Charter Revision in the Empire State*, 247.

12. *Proceedings, 1967*, vol. II, Record, Part I, 528.

13. *Proceedings, 1967*, vol. II, Record, Part I, 542.

14. *Proceedings, 1967*, vol. II, Record, Part I, 544–46. Frank Graham writes, "It was not common knowledge even among preservationists that her speech like that of most of the other Democratic delegates pleading for the Forever Wild clause, had been written by David Sive" (Graham, *The Adirondack Park*, 217). Sive was an ardent defender of Article 14, chair of the Atlantic Chapter of the Sierra Club, and head of the Democratic staff supporting the convention's Committee on Natural Resources and Agriculture. Graham, 214–17, describes Sive's extensive efforts at the convention to retain Article 14 in precisely its original language, but for none of his assertions, including that Sive was the author of Robinson's remarks, does Graham provide any documentation. Graham's *The Adirondack Park* is an essential source on the history of the Adirondacks, but its identification of sources is iffy. For more on Sive and his role at the 1967 convention, see Edmondson, *A Wild Idea*, 52–53, 57–59. Edmondson interviewed Sive before he died, and Sive confirmed that he had written Robinson's speech. Edmondson clarified this to me in an email, May 5, 2022. So Graham, notwithstanding his casual system of documentation, had it right.

15. *Proceedings, 1967*, vol. II, Record, Part I, 546–47.

16. See Dullea, *Charter Revision in the Empire State*, 333–48.

22. Balsam Lake

1. See my *Forever Wild*, 159–63, and *Contested Terrain*, 168–71. See also Graham, *The Adirondack Park*, 247–49, and Edmondson, *A Wild Idea*, 215–21.

2. Background information is from the final decision handed down by the Appellate Division, Third Department, in *Balsam Lake Anglers Club v. Dept. of Envtl. Conservation*, December 30, 1993, 199 A.D.2d 852 (N.Y. App. Div. 1993), 605 N.Y.S.2d 795, https://casetext.com/case/matter-of-balsam-lake-v-dept-of-envtl, hereafter *Balsam Lake*.

3. Quotation is from the decision issued by the Supreme Court of Ulster County in *Balsam Lake Anglers Club v. Department of Environmental Conservation*, December 30, 1991, https://casetext.com/case/balsam-anglers-club-v-dec?.

4. https://casetext.com/case/matter-of-balsam-lake-v-dept-of-envtl.

5. This distillation of the two decisions and how they measure acceptable and unacceptable levels of tree cutting comes from an assessment of these two cases and *Protect* by Peter Bauer, available on the Protect website at https://www.protectadks.org/the-meaning-of-the-2021-new-york-constitution-article-14-forever-wild-decision/.

23. Wilderness and Constitutions

1. For a recent and riveting account of these years, see Edmondson, *A Wild Idea*. See also Graham, *The Adirondack Park*, 230–53.

2. The official name of this organization is Protect the Adirondacks! Inc., with an exclamation mark. Its roots lie in a merger between the Association for the Protection of the Adirondacks, established in 1902, and the Residents' Committee to Protect the Adirondacks, established in 1990. In 2009 both of these groups were dealing with insufficient funding, declining membership, and leadership issues, so they joined forces and soon thereafter hired Peter Bauer as executive director. I find the exclamation point to be distracting when the name is repeated frequently as it is here and choose not to use it. In this book, as noted earlier, the organization is Protect, the litigation is *Protect*.

3. Cole, "Originalism's Charade," 18. This article is a lengthy and insightful review of Chemerinsky, *Worse than Nothing*.

24. *Protect the Adirondacks! Inc. v. New York State Department of Environmental Conservation and Adirondack Park Agency*

1. John Warren published a series of articles on the history of regional snowmobile use in the *Adirondack Almanack* (which he founded) in 2007 (October 1, 3, 5, 8, and 10). These are collected in his *Historic Tales from the Adirondack Almanack*, 56–70.

2. See Edmondson, *A Wild Idea*, 13–14. His interview with Coggeshall and Newhouse, a transcript of which Edmondson shared with me and is in my files, took place on October 23 and 24, 2003. Email to me from David Gibson, March 17, 2024, in my files.

3. Personal correspondence with Peter Paine, February 2 and 3, 2023, now in my files.

4. See Edmondson, *A Wild Idea*, 90, 211–21; Terrie, *Forever Wild*, 159–63.

5. Personal correspondence with Peter Bauer, January 27–28, 2024; conversation with Dave Gibson, March 15, 2024.

6. The key documents for this section are the decision from the Supreme Court (https://www.protectadks.org/wp-content/uploads/2017/12/Article-14-Decision .12.1.2017.pdf), the decision from the Appellate Division (https://casetext.com/case /protect-the-adirondacks-inc-v-ny-state-dept-of-envtl-conservation-12), and the final decision issued by the New York Court of Appeals (https://casetext.com/case/protect -the-adirondacks-inc-v-ny-state-dept-of-envtl-conservation-15). Quotations from these decisions provide nearly all the evidence in this section, and I clarify as I use them which court I am referring to.

7. Spector, "Cuomo"; Esch, "Adirondack Land Decisions Draw Praise, Criticism"; Smith Dedam, "Adirondack Challenge."

8. Anthony DePalma, "Conservancy Buys Large Area of Adirondack Wilderness," *New York Times* (June 19, 2007).

9. Warren, "New York State Acquires 69,000 Acres From Conservancy," *Adirondack Almanack* (August 5, 2012); Phil Brown, "State to Buy 69,000 Acres," *Adirondack Explorer* (Aug. 20, 2012), https://www.adirondackexplorer.org/stories/state-to -buy-69000-acres.

10. See McMartin, *Perspectives on the Adirondacks*, 187.

11. Interviews with Peter Bauer, January 25 and March 15, 2024.

12. Article 14, Section 5, states, "A violation of any of the provisions of this article may be restrained at the suit of the people or, with the consent of the supreme court in appellate division, on notice to the attorney general at the suit of the citizen." This was one of the provisions added to then Article 7 at the 1938 Constitutional Convention. I am grateful to Claudia Braymer for clarifying this.

13. For clarification of the mechanics of the litigation, I am grateful to Peter Bauer and Claudia Braymer.

14. The other was in a case involving public access to historic canoe routes, *Friends of Thayer Lake LLC v. Brown*. In that case, as with *Protect*, the lawyer prepping me and then quizzing me on the stand was John Caffry.

15. Judge Connolly's decision was issued December 1, 2017.

16. This decision was issued July 3, 2019.

17. ADK's argument, that if the Court of Appeals ruled in favor of Protect, the State would not be able to build or improve foot trails anywhere on the Forest Preserve, struck me as stunningly unpersuasive. Not only was it wrong—a misreading of the record, including the decisions in *MacDonald* and *Balsam Lake*, and a radical misunderstanding of the obvious goal of Protect in initiating the suit in the first place—it was also a 180-degree pivot from nearly a century of ADK's history. My dismay at ADK's willingness to argue on the side of the State led to a testy exchange

between Michael Barrett, executive director at ADK, and me, conducted first in private and then publicly at the *Adirondack Almanack*, where I made my case and he made his. See my "Has the Adirondack Mountain Club Lost Its Way?," *Adirondack Almanack* (March 8, 2021), https://www.adirondackalmanack.com/2021/03/commentary -has-the-adirondack-mountain-club-lost-its-way.html, and Michael Barrett, "ADK's Support of Sustainable Trails," *Adirondack Almanack* (March 10, 2021), https://www .adirondackalmanack.com/2021/03/adks-support-of-sustainable-trails.html. ADK's refusal to discuss in any way how it made the decision to file its amicus brief led me to leave the club, having first joined it over fifty years earlier. ADK used to be a member-run, grassroots organization. Now, after major revision of its governance documents that limits the participation of members in critical Club decisions and rewrites the process by which the Board of Directors is constituted, it has become a brand, a business; see all the changes and the new bylaws, which nowhere include the word "wilderness," in *Adirondac* (September–October 2020): 24–29; see also Zachary Matson, "ADK's Next 100 Years," *Adirondack Explorer* (January–February 2022): 27.

25. The Decision in *Protect*

1. In addition to the Court's decision, see Craig, "Court of Appeals"; Silvarole, "Plan for Snowmobile Trails."

2. The occasion was an all-day seminar commemorating the fiftieth anniversary of the establishment of the Adirondack Park Agency, hosted by the Adirondack Experience (a.k.a. the Adirondack Museum). In an op-ed in the *Explorer*, Martens characterized what he said to the towns as a "commitment": "The New Law of the Land" (July/August 2021), 57.

3. James Odato, "Trail Promise Melts," *Adirondack Explorer* (July/August 2021): 14–17.

4. Gwendolyn Craig, "Judgment May End Snowmobile Trail Litigation," *Adirondack Explorer* (October 5, 2023), https://www.adirondackexplorer.org/stories/judgment -may-end-snowmobile-trail-litigation.

5. See Peter Bauer, "DEC-APA Aim To Violate Forever Wild Clause Again," *Adirondack Almanack* (October 10, 2024), https://www.adirondackalmanack.com/2024/10 /dec-apa-aim-to-violate-forever-wild-clause-again.html.

26. Wilderness in the Sewards

1. Marshall, "Approach to the Mountains," 105–8.

2. This narrative of a hike in the Sewards was originally published, in somewhat different form, as Terrie, "Wilderness Guaranteed," 44–49.

3. White, *Adirondack Country*, 164.

4. Colvin, *Ascent and Barometrical Measurement of Mount Seward*.

5. Colvin, *Ascent and Barometrical Measurement of Mount Seward*, 174.

6. Marshall, *The High Peaks of the Adirondacks*, 27–28.

7. Goodwin, "The Forty-Six Peaks," 64.

8. Swan, "The Seward Range," 289–92.

9. US Forest Service, *Service Bulletin* 12, no. 35 (August 17, 1928): 5–6. This newsletter, marked "Contents Confidential" on the first page, apparently served more or less as an ongoing interoffice memo. My copy was located for me at the University of California, Berkeley, by a helpful librarian, name unknown to me, at Bowling Green State University, where I had requested it via interlibrary loan.

10. See Sutter's discussion of Marshall's "The Wilderness as a Minority Right," *Driven Wild*, 87, 208–9.

27. Whose Wilderness?

1. See, among many others, Hill, "In White Adirondacks"; Aaron Marbone, "Family: Racism Drove Student Out of School," *Adirondack Daily Enterprise* (January 6, 2023); Sandreczki, "Are the Adirondacks Really Ready for Everyone?"

2. Murray, *Adventures in the Wilderness*. Anyone interested in the whole affair of Murray and Murray's Fools should track down the 1970 reprint of *Adventures in the Wilderness*, published by the Adirondack Museum and Syracuse University Press, with an indispensable, thoroughly researched introduction by the late Warder H. Cadbury. All studies of Murray should start with Cadbury's Introduction, which is my source for the various quotations that follow. Understanding citations from Cadbury can be tricky: the pagination in the reprint uses Arabic numbers in italics. See also Strauss, "Toward a Consumer Culture," 270–86. Some of this chapter was originally published as "'Murray's Fools' at 150," in the November/December 2019 issue of the *Adirondack Explorer*.

3. Eight letters over the pen name "Wachusett" appeared in the *Boston Daily Advertiser* between July 17 and 30, 1869: Cadbury, Introduction, *40*.

4. Thorpe, "The Abuses of the Backwoods," *564–65*, quoted in Cadbury, Introduction, *48*. See also Thorpe, "A Visit to 'John Brown's Tract,'" 160–78.

5. Kate Field, "Among the Adirondacks: Murray's Fools—A Plain Talk about the Wilderness," *New-York Tribune* (August 12, 1869), quoted in Cadbury, Introduction, *48*; Thorpe, "The Abuses of the Backwoods."

6. *Forest and Stream*, vol. 31 (August 9, 1888): 46.

7. Terrie, *Contested Terrain*, 120.

8. For example, see *Adirondack Almanack* (May 3, 2023), https://www.adirondack almanack.com/2023/05/rangers-assist-cold-wet-unprepared-hikers-at-lake-colden -outpost.html#more-209985.

9. Otis, *Rural Indigenousness*; Miller, *This Radical Land*; Svenson, *Blacks in the Adirondacks*; Godine, *The Black Woods*.

10. Stoddard, *The Adirondacks Illustrated*, 68–69. I found Stoddard's anecdote discussed in Godine, *The Black Woods*, 295–96.

11. Sutter, *Driven Wild*, 194–95.

12. See MacMillan, "Robert Marshall."

13. Young, *Heading Out*, 172–207.

14. "All are welcome where they once were not" (February 1, 2022), https://nystateparks.blog/2022/02/01/all-are-welcome-where-they-once-were-not/.

15. This and similar incidents were reported by Mann, "Top African American Environmental Leader."

16. Russell, "Truck with Confederate Flag."

28. The Constitution, 2038

1. Nearing, "New York Could Lose Half"; Alan Wechsler, "Requiem for Ice," *Adirondack Explorer* (March/April 2024).

2. For a thoughtful synopsis of how climate change challenges any traditional concept of wilderness, see Christopher Solomon, "Rethinking the Wild," *New York Times* (July 6, 2014), https://www.nytimes.com/2014/07/06/opinion/sunday/the-wilderness-act-is-facing-a-midlife-crisis.html.

3. McKibben, *The End of Nature*; Stager, *Deep Future*; Kolbert, *Field Notes from a Catastrophe*; and Mann, *The New Climate War*.

4. This study was released and posted online, February 2024, https://nysclimateimpacts.org. The information developed in this paragraph comes from this report.

5. Daryl McGrath, "Birds Visit New Areas or Vanish from the Adirondacks," *Adirondack Explorer* (October 10, 2023), https://www.adirondackexplorer.org/stories/birds-visit-new-areas-or-vanish-from-the-adirondacks.

6. https://nysclimateimpacts.org.

7. Jenkins, *Climate Change in the Adirondacks*, 25–67. See especially the graphic on p. 29.

8. United Nations Environment Programme, *Spreading like Wildfire: The Rising Threat of Extraordinary Landscape Fires* (Nairobi, 2022). PDF available at https://www.unep.org/resources/report/spreading-wildfire-rising-threat-extraordinary-landscape-fires.

9. Curth, "Climate Change," 101–4. See also Chloe Bennett, "It's Complicated: Climate Change and Wildfires in the Adirondacks," *Adirondack Explorer* (July/Aug. 2023): 22–23.

10. See Semerad, "Article XIV," 10, where constitutional lawyer Semerad notes that the decision of the attorney general to permit not only removal of downed timber but also its sale was "an extremely liberal construction of the constitution." See also Edmondson, *A Wild Idea*, 15–16, 24–26. I suspect that if the blowdown happened

today, any attempt by the state to undertake or permit salvage operations would be challenged in court. After a violent windstorm in July 1995, there were similar fears about fire and calls for salvage operations on state land, but they went nowhere. And no fires occurred.

11. Jenkins, *Climate Change in the Adirondacks*, 130–44. One significant variable concerns the age of the forest. It might seem reasonable to conclude that a younger forest with rapidly growing trees might constitute a better carbon storage system than an old-growth forest. Evidence suggests that that is not the case, especially when all carbon inputs and outputs involved in logging and manufacturing are considered. See Hudiburg et al., "Meeting GHG Reduction Standards." See also Climate and Applied Forest Research Institute, "New York Forest Carbon Assessment" (June 2023), https://www.esf.edu/cafri-ny/documents/cafri-report-2023.pdf.

12. See Chloe Bennett, "Seeking Refuge in the Adirondacks," *Adirondack Explorer* (March/April 2024): 12–15.

13. In November of 2023 Protect the Adirondacks released a detailed study of "30x30"; see Bauer, Braymer, McMartin Long, and Signell, *20% in 2023*. See also Bauer and Braymer, "Will the State of New York Achieve Its Goal?," 109–21.

14. "The Forest Preserve's Expanding Size and Role," *Adirondack Explorer* (November/December 2023), 40–41. Online comments at https://www.adirondack explorer.org/stories/the-forest-preserves-expanding-size-and-role#comments. The argument that the state can't take care of what it already has is nonsense: the Forest Preserve takes care of itself. What the state does have difficulty managing is people, especially as DEC budgets shrink and the personnel needed to educate and oversee growing numbers of New Yorkers who want to experience the wilderness simply aren't there. The problem is not state ownership but some of the people who use the Forest Preserve; they litter and get lost and injured.

15. Jesse McKinley and Brad Plumer, "New York to Approve One of the World's Most Ambitious Climate Plans," *New York Times* (June 18, 2019), https://www.ny times.com/2019/06/18/nyregion/greenhouse-gases-ny.html.

Bibliography

Adirondac.

Adirondack Daily Enterprise.

Adirondack Explorer.

Adirondack Forty-Sixers. *The Adirondack High Peaks.* Adirondack Forty-Sixers, 1970.

Adirondack Forty-Sixers. *Heaven Up-h'sted-ness.* Adirondack Forty-Sixers, 2011.

Adirondack Forty-Sixers. *Of the Summits, Of the Forests.* Adirondack Forty-Sixers, 1991.

Adirondack Mountain Club. *Adirondack Bibliography.* Adirondack Mountain Club, 1958.

Adirondack Mountain Club. *Adirondack Bibliography Supplement 1956–1965.* Adirondack Museum, 1973.

Adirondack Mountain Club. *Guide to Adirondack Trails: High Peak Region and Northville-Placid Trail,* 7th ed. Adirondack Mountain Club, 1962.

Adirondack Park Regional Assessment Project. "*The Adirondack Park*: Seeking Balance." Adirondack Park Regional Assessment Project, 2014.

The Atlantic.

Bartram, William. *Travels.* Viking Penguin, 1988.

Bauer, Peter, and Claudia Braymer. "Will the State of New York Achieve Its Goal of Protecting 30% of the State's Land and Inland Waters by 2030?" *Environmental Law In New York* 35 (August 2024): 109–21.

Bauer, Peter, Claudia Braymer, James McMartin Long, and Stephen Signell. *20% in 2023: An Assessment of the New York State 30 by 30 Act.* Protect the Adirondacks, 2023.

Berger, Peter L., and Thomas Luckmann. *The Social Construction of Reality: A Treatise in the Sociology of Knowledge.* Anchor Books, 1966.

Bradford, William. *Of Plymouth Plantation, 1620–1647*. The Modern Library, 1981.

Bonehill, Ralph. *Pioneer Boys of the Great Northwest, Or With Lewis and Clark across the Rockies*. Grosset and Dunlap, 1904.

Cadbury, Warder H. Introduction to *Adventures in the Wilderness; or, Camp-Life in the Adirondacks*, by William H. H. Murray. 1869. Reprint, Adirondack Museum and Syracuse University Press, 1970.

Callicott, J. Baird, and Michael P. Nelson, eds. *The Great New Wilderness Debate: An Expansive Collection of Writings Defining Wilderness from John Muir to Gary Snyder*. Univ. of Georgia Press, 1998.

Carson, Russell. *Peaks and People of the Adirondacks*. Doubleday, Page, 1927; Adirondack Mountain Club, 1972.

Chamberlin, Silas. *On the Trail: A History of American Hiking*. Yale Univ. Press, 2016.

Chemerinsky, Erwin. *Worse than Nothing: The Dangerous Fallacy of Originalism*. Yale Univ. Press, 2022.

Cohen, Michael P. *The Pathless Way: John Muir and American Wilderness*. Univ. of Wisconsin Press, 1984.

Cole, David. "Originalism's Charade." *New York Review of Books*, November 24, 2022.

Colvin, Verplanck. *Ascent and Barometrical Measurement of Mount Seward*. Argus, 1872.

Colvin, Verplanck. *Report of the Superintendent of the State Land Survey*. Senate Document 42, 1896; Wynkoop Hallenbeck Crawford, 1896.

Colvin, Verplanck. *Seventh Annual Report on the Progress of the Topographical Survey of the Adirondack Region of New York, to the year 1879, Containing the Condensed Reports for the Years 1874–75–76–77 and '78*. Assembly Document 87, 1879; Weed, Parsons, 1880.

Conaway, James. *America's Library: The Story of the Library of Congress, 1800–2000*. Yale Univ. Press, 2000.

Craig, Gwendolyn. "Court of Appeals Bars Cutting Trees for Snowmobile Trails." *Albany Times-Union*, March 4, 2021.

Cronon, William. *Changes in the Land: Indians, Colonists, and the Ecology of New England*. Hill and Wang, 1983.

Cronon, William. "The Trouble with Wilderness." *New York Times Magazine*, August 13, 1995.

Cronon, William. "The Trouble with Wilderness." In *Uncommon Ground: Toward Reinventing Nature*, edited by William Cronon. W. W. Norton, 1995.

Curth, Louis. "Climate Change and the Myth of the Adirondack Asbestos Forest." *Adirondack Journal of Environmental Studies* 25 (2022): 101–4.

Denevan, William M. "The 'Pristine Myth' Revisited." *Geographical Review* 101 (October 2011): 576–91.

Donaldson, Alfred Lee. *A History of the Adirondacks*. 2 vols. Century, 1921.

Dullea, Henrik N. *Charter Revision in the Empire State: The Politics of New York's 1967 Constitutional Convention*. The Rockefeller Institute, 1997.

Edmondson, Brad. *A Wild Idea: How the Environmental Movement Tamed the Adirondacks*. Cornell Univ. Press, 2021.

Esch, Mar. "Adirondack Land Decisions Draw Praise, Criticism." *Associated Press*, May 7, 2016, https://www.pressconnects.com/story/news/local/new-york/2016/05/07/adirondack-decisions-draw-praise-criticism/84070378/.

Fox, Stephen. *John Muir and His Legacy: The American Conservation Movement*. Little, Brown, 1981.

Galie, Peter J. *Ordered Liberty: A Constitutional History of New York*. Fordham Univ. Press, 1996.

Glover, James. *A Wilderness Original: The Life of Bob Marshall*. The Mountaineers, 1986.

Godine, Amy. *The Black Woods: Pursuing Racial Justice on the Adirondack Frontier*. Cornell Univ. Press, 2023.

Goodwin, Jim. "The Forty-Six Peaks." In *Of the Summits, Of the Forests*, edited by Adirondack Forty-Sixers. Adirondack Forty-Sixers, 1991.

Gordon-Reed, Annette. *The Hemingses of Monticello: An American Family*. W. W. Norton, 2008.

Graham, Frank. *The Adirondack Park: A Political History*. Alfred A Knopf, 1978.

Halper, Louise A. "'A Rich Man's Paradise': Constitutional Preservation of New York State's Adirondack Forest, A Centenary Consideration." *Ecology Law Quarterly* 19, no. 2 (1992): 193–267.

Hämäläinen, Pekka. *Indigenous Continent: The Epic Contest for North America*. New York: W. W. Norton, 2022.

Hammond, Samuel H. *Hills, Lakes and Forest Streams*. J. C. Derby, 1854.

Harper, Kyle. *The Fate of Rome: Climate, Disease, and the End of an Empire*. Princeton Univ. Press, 2017.

Harvey, Mark. *Wilderness Forever: Howard Zahniser and the Path to the Wilderness Act.* Univ. of Washington Press, 2005.

Headley, Joel T. *The Adirondack: or, Life in the Woods.* Baker and Scribner, 1849.

Headley, Joel T. *The Great Riots of New York.* E. G. Treat, 1873.

Headley, Joel T. *Washington and His Generals.* Baker and Scribner, 1847.

Hill, Michael. "In White Adirondacks, Racism May be Toughest Hill to Climb." *Washington Post*, September 14, 2020.

Hoffman, Charles Fenno. *Wild Scenes in the Forest and Prairie.* W. H. Colyer, 1843.

Hopsicker, Peter. "Legalizing the 1932 Lake Placid Olympic Bobrun: A Test of the Adirondack Wilderness Culture." *Olympika: The International Journal of Olympic Studies* 18 (2009): 99–120.

hooks, bell. *Belonging: A Culture of Place.* Routledge, 2009.

Hudiburg, Tara, et al. "Meeting GHG Reduction Standards Requires Accounting for All Forest Sector Emissions." *Environmental Research Letters* 14, no. 9 (2019) 095005, DOI:10.1088/1748-9326/ab28bb.

Jacoby, Karl. *Crimes Against Nature: Squatters, Poachers, Thieves, and the Hidden History of American Conservation.* Univ. of California Press, 2001.

James, N. D. G. *Historical Dictionary of Forestry and Woodland Terms.* Blackwell, 1991.

Jenkins, Jerry. *Climate Change in the Adirondacks: The Path to Sustainability.* Cornell Univ. Press and the Wildlife Conservation Society, 2010.

Jennings, Francis. *The Invasion of America: Indians, Colonialism, and the Cant of Conquest.* The Omohundro Institute of Early American History and Culture and Univ. of North Carolina Press, 1975.

Johnson, Deidre. *Edward Stratemeyer and the Stratemeyer Syndicate.* Twayne, 1993.

Kaye, Roger. "Conservation Controversy: Robert Marshall's Fight to Keep Northern Alaska Wild." *Heartland Magazine*, n.d.

Kolbert, Elizabeth. *Field Notes from a Catastrophe.* Bloomsbury, 2006.

Kolodny, Annette. *The Land Before Her: Fantasy and Experience of the American Frontiers, 1630–1860.* Univ. of North Carolina Press, 1984.

Kolodny, Annette. *The Lay of the Land: Metaphor as Experience and History in American Life and Letters.* Univ. of North Carolina Press, 1975.

Kuklick, Bruce. "Myth and Symbol in American Studies." *American Quarterly* 24 (1972): 435–50.

Lears, T. J. Jackson. *No Place of Grace: Antimodernism and the Transformation of American Culture, 1880–1920*. Pantheon, 1981.

Lincoln, Charles Zebina. *The Constitutional History of New York, from the Beginning of the Colonial Period to the year 1905, 5 vols.* The Lawyers Cooperative Publishing Co., 1905, https://nysl.ptfs.com/#!/s?a=c&q=*&type=16&criteria=field11%3D1337955&b=0.

Long, James McMartin, and Peter Bauer. *The Adirondack Park and Rural America: Economic and Population Trends, 1970–2010*. Protect the Adirondacks! Inc., 2019.

Longstreth, T. Morris. *The Adirondacks*. Century, 1917.

MacMillan, Ian. "Robert Marshall: The Red and Green 'Map Maker of Utopia.'" PhD dissertation, Queen's Univ., 2002. ProQuest (NQ99843).

Mann, Brian. "Top African American Environmental Leader Faces Racial Incident in Adirondacks." *North Country Public Radio*, October 31, 2016, https://www.northcountrypublicradio.org/news/story/32831/20161031/top-african-american-environmental-leader-faces-racial-incident-in-adirondacks%E2%80%AC.

Mann, Charles C. *1491: New Revelations of the Americas Before Columbus*. Vintage Books, 2005.

Mann, Michael. *The New Climate War*. Public Affairs, 2021.

Marsh, George Perkins. *Man and Nature: or Physical Geography as Modified by Human Action*. Charles Scribner, 1864.

Marshall, George. "Approach to the Mountains." *Adirondac* (March–April 1955). In *The Adirondack Reader*, edited by Paul Jamieson, 105–8. MacMillan, 1964.

Marshall, Robert. *Alaska Wilderness: Exploring the Central Brooks Range*. Univ. of California Press, 1956.

Marshall, Robert. *Arctic Village*. Literary Guild, 1933.

Marshall, Robert. *The High Peaks of the Adirondacks*. Adirondack Mountain Club, 1922.

Marshall, Robert. "The Problem of the Wilderness." *Scientific Monthly* 30 (February 1930): 141–48.

Marx, Leo. *The Machine in the Garden: Technology and the Pastoral Ideal in America*. Oxford Univ. Press, 1964.

McKibben, Bill. *The End of Nature*. Random House, 1989.

McMartin, Barbara. *Perspectives on the Adirondacks: A Thirty-Year Struggle by People Protecting their Treasure*. Syracuse Univ. Press, 2002.

Meine, Curt. *Aldo Leopold: His Life and Work*. Univ. of Wisconsin Press, 1988.

Miller, Daegan Ryan. *This Radical Land: A Natural History of American Dissent*. Univ. of Chicago Press, 2018.

Miller, Daegan Ryan. "Witness Tree: Landscape and Dissent in the Nineteenth-Century United States." PhD Dissertation, Cornell Univ., 2013. ProQuest (3571207).

Morison, Samuel Eliot. *Samuel de Champlain: Father of New France*. Little, Brown, 1972.

Morris, Wade Jr. "Forcing Progress: The Struggle to Integrate Southern Episcopal Schools." MA thesis, Georgetown Univ., 2009. ProQuest (1462699).

Murray, William H. H. *Adventures in the Wilderness; or, Camp-Life in the Adirondacks*. Fields, Osgood, 1869. Reprint, Adirondack Museum and Syracuse University Press, 1970.

Nash, Roderick. "*Forever Wild: Environmental Aesthetics and the Adirondack Forest Preserve* (Book Review)." *The American Historical Review* 91, no. 2 (April 1986): 478–79.

Nash, Roderick. *Wilderness and the American Mind*. Yale Univ. Press, 1967.

Nearing, Brian. "New York Could Lose Half Its Ski Season by 2050." *Albany Times Union*, January 15, 2018.

Nelson, Michael P., and J. Baird Callicott, eds. *The Wilderness Debate Rages On*. Univ. of Georgia Press, 2008.

New York State Conservationist, May–June 1985.

"The New York Natural History Survey in the Adirondack Wilderness, 1836–1840." *Journal of the Early Republic* 3 (July 1983): 185–206.

New York Times.

Nobles, Gregory H. *John James Audubon: The Nature of the American Woodsman*. Univ. of Pennsylvania Press, 2017.

Novak, Barbara. *American Landscape and Painting, 1825–1875*. Oxford Univ. Press, 1995.

Nussbaum, Martha C. "A Peopled Wilderness." *New York Review of Books*, December 8, 2022, 21–24.

O'Rourke, Vernon, and Douglas W. Campbell. *Constitution-Making in a Democracy: Theory and Practice in New York State*. Johns Hopkins Univ. Press, 1943.

Otis, Melissa. *Rural Indigenousness: A History of Iroquoian and Algonquian Peoples in the Adirondacks*. Syracuse Univ. Press, 2018.

Ouderkirk, Wayne. "On Wilderness and People: A View from Mt. Marcy." In *The Wilderness Debate Rages On*, edited by Michael P. Nelson and J. Baird Callicott, 435–60. Univ. of Georgia Press, 2008.

Pownall, Thomas. *A Topographical Description of the Dominions of the United States of America*. Edited by Lois Mulkearn. Univ. of Pittsburgh Press, 1949.

Proceedings of the Constitutional Convention of the State of New York, April fourth to September twenty-sixth, 1967. 12 vols., https://nysl.ptfs.com/#!/s?a=c&q=*&type=16&criteria=field11%3D19122&b=0.

Pyne, Stephen J. *Fire in America: A Cultural History of Wildland and Rural Fire*. Princeton Univ. Press, 1982.

"Report of the Citizens' Advisory Task Force on Open Space to the Adirondack Park Agency." Bound typescript, April 18, 1980.

Revised Record of the Constitutional Convention of the State of New York, April Sixth to September Tenth 1915, 4 vols. J. B. Lyon, 1916, https://catalog.hathitrust.org/Record/001156630.

Revised Record of the Constitutional Convention of the State of New York, May 8, 1894, to September 29, 1894, 5 vols. Argus Co., 1900, https://www.google.com/books/edition/Revised_Record_of_the_Constitutional_Con/SG00AQAAMAAJ?hl=en&gbpv=1&printsec=frontcover.

Reynolds, Janet. "Early Nomads or Occupants of the Adirondacks? Examining the Ancestral Home and Range of Indigenous People." *Adirondack Explorer*, July–August 2023.

Rosevear, Francis. *Colvin in the Adirondacks: a Chronology*. North Country Books, 1992.

Runte, Alfred. *National Parks: The American Experience*. Univ. of Nebraska Press, 1979.

Russell, Emily. "Truck with Confederate Flag Sparks Carnival Controversy in Saranac Lake." *North Country Public Radio*, February 15, 2024, https://www.northcountrypublicradio.org/news/story/49300/20240215/truck-with-confederate-flag-sparks-carnival-controversy-in-saranac-lake.

Ryan, Jeffrey H. *This Land was Saved for You and Me: How Gifford Pinchot, Frederick Law Olmsted, and a Band of Foresters Rescued America's Public Lands*. Stackpole Books, 2022.

Sandreczki, Monica. "Are the Adirondacks Really Ready for Everyone? Community Groups Get Real about Racial, Environmental Justice." *North Country Public Radio*, May 26, 2022, https://www.northcountry

publicradio.org/news/story/45957/20220526/are-the-adirondacks-really
-for-everyone-community-groups-get-real-about-racial-environmental
-justice.

Schaefer, Paul, ed. *Adirondack Explorations: Nature Writings of Verplanck Colvin*. Syracuse Univ. Press, 1997.

Schrepfer, Susan R. *Nature's Altars: Mountains, Gender, and American Environmentalism*. Univ. of Kansas Press, 2005.

Sears, John F. *Sacred Places: American Tourist Attractions in the Nineteenth Century*. Oxford Univ. Press, 1989.

Seidule, Ty. *Robert E. Lee and Me: A Southerner's Reckoning with the Myth of the Lost Cause*. St. Martins, 2021.

Semerad. Ralph. "Article XIV." *The Future of the Adirondacks: The Reports of the Temporary Study Commission on the Future of the Adirondacks*. Adirondack Museum, 1971.

Silvarole, Georgie. "Plan for Snowmobile Trails through Adirondack Park Ruled Unconstitutional. Here's Why." *Rochester Democrat & Chronicle*, May 2, 2021.

Silver, M. M. *Louis Marshall and the Rise of Jewish Ethnicity in America*. Syracuse Univ. Press, 2013.

Slotkin, Richard. *Regeneration Through Violence: The Mythology of the American Frontier, 1600–1860*. Wesleyan Univ. Press, 1973.

Smith, Henry Nash. *Virgin Land: The American West as Symbol and Myth*. Harvard Univ. Press, 1950.

Smith Dedam, Kim. "Adirondack Challenge Spreads Positive Regional Image." *Plattsburgh Press Republican*, March 10, 2014.

Spector, Joseph. "Cuomo: Come to NY for the Snow, Stay for the Free Snowmobiling." *Lohud*, February 20, 2015, https://www.lohud.com/story/news/politics/politics-on-the-hudson/2015/02/20/cuomo-come-to-ny-for-the-snow-stay-for-the-free-snowmobiling/23754469/.

Spence, Mark David. *Dispossessing the Wilderness: Indian Removal and the Making of the National Parks*, rev. ed. Oxford Univ. Press, 2000.

Stager, Curt. *Deep Future: The Next 100,000 Years of Life on Earth*. St. Martin's, 2010.

Stager, Curt. "Hidden Heritage." *Adirondack Life* 48 (June 2017).

Stoddard, Seneca Ray. *The Adirondacks Illustrated*. Weed, Parsons, 1874.

Strauss, David. "Toward a Consumer Culture: 'Adirondack Murray' and the Wilderness Vacation." *American Quarterly* 39 (Summer 1987).

Street, Alfred B. *The Indian Pass*. Hurd and Houghton, 1869.

Street, Alfred B. *Woods and Waters; Or the Saranac and Racket*. M. Doolady, 1860.

Sutter, Paul. *Driven Wild: How the Fight Against Automobiles Launched the Modern Wilderness Movement*. Univ. of Washington Press, 2002.

Svenson, Sally. *Blacks in the Adirondacks: A History*. Syracuse Univ. Press, 2017.

Swan, John Sharp Jr. "The Seward Range." In *Heaven Up-h'sted-ness*, edited by Adirondack Forty-Sixers. Adirondack Forty-Sixers, 2011.

Terrie, Philip G. "The Adirondack Forest Preserve: The Irony of Forever Wild." *New York History* 62 (July 1981): 261–88.

Terrie, Philip G. *Contested Terrain: A New History of Nature and People in the Adirondacks*. 2nd ed. Syracuse Univ. Press, 2008.

Terrie, Philip G. *Forever Wild: A Cultural History of Wilderness in the Adirondacks*. Syracuse Univ. Press, 1994.

Terrie, Philip G. *Forever Wild: Environmental Aesthetics and the Adirondack Forest Preserve*. Temple Univ. Press, 1985.

Terrie, Philip G. "'The Grandest Private Park': Forestry and Land Management." In *The Adirondack League Club, 1890–1990*, edited by Edward Comstock, 73–111. Adirondack League Club, 1990.

Terrie, Philip G. "The New York Natural History Survey in the Adirondack Wilderness, 1836–1840." *Journal of the Early Republic* 3 (July 1983): 185–206.

Terrie, Philip G. "Romantic Travelers in the Adirondack Wilderness." *American Studies* 24 (Fall 1983): 59–75.

Terrie, Philip G. *Seeing the Forest: Reviews, Musings, and Opinions from an Adirondack Historian*. Lost Pond Press/*Adirondack Explorer*, 2017.

Terrie, Philip G. "Urban Man Confronts the Wilderness: The Nineteenth-Century Sportsman in the Adirondacks." *Journal of Sport History* 5 (Winter 1978): 7–20.

Terrie, Philip G. "Verplanck Colvin, Adirondack Surveyor: A Case Study of the Late Nineteenth-Century Response to Wilderness." *Environmental Review* 7 (Fall 1983): 275.

Terrie, Philip G. *Wildlife and Wilderness: A History of Adirondack Mammals*. Purple Mountain Press, 1993.

Terrie, Philip G. "Wilderness Guaranteed: Celebrating a Century of Forever Wild." *Adirondack Life* 55 (November–December 1994): 44–49.

Thorpe, Thomas Bangs. "The Abuses of the Backwoods." *Appleton's Journal* (December 18, 1869): 564–65.

Thorpe, Thomas Bangs. "A Visit to John Brown's Tract." *Harper's New Monthly Magazine* 19 (June–November 1859): 160–78.

Tuan, Yi-Fu. *Topophilia: A Study of Environmental Perception, Attitudes, and Values*. Columbia Univ. Press, 1974.

Tyler, Anthony O., and Michael Wilson, eds. *Inside the Blue Line: Essays on Adirondack Environments*. Potsdam College Press, 2009.

VanValkenburgh, Norman J. *The Adirondack Forest Preserve: A Narrative of the Evolution of the Adirondack Forest Preserve of New York State*. Adirondack Museum, 1979.

Ward, John William. *Andrew Jackson: Symbol for an Age*. Oxford Univ. Press, 1955.

Warren, John. *Historic Tales from the Adirondack Almanack*. The History Press, 2009.

Webster's International Dictionary of the English Language. G & C Merriam Co., 1892.

Welch, Douglas B., comp. *Adirondack Books, 1962–1992: An Annotated Bibliography*. North Country Books, 1994.

White, Dan. *Under the Stars: How America Fell in Love with Camping*. Henry Holt, 2016.

White, Richard. "'Are You an Environmentalist, or Do You Work for a Living?': Work and Nature." William Cronon, ed., *Uncommon Ground: Toward Reinventing Nature*. W. W. Norton, 1995.

White, William Chapman. *Adirondack Country*. Alfred A. Knopf, 1967.

Wise, Gene. "'Paradigm Dramas' in American Studies and Institutional History of the Movement." *American Quarterly* 31 (1977): 293–337.

Worster, Donald. *Nature's Economy: A History of Ecological Ideas*. Cambridge Univ. Press, 1977.

Young, Terence. *Heading Out: A History of American Camping*. Cornell Univ. Press, 2017.

Index

Philip G. Terrie has been exploring the story of wilderness in the Adirondacks, both in the archives and deep in the woods, for over fifty years. Early in his career, he worked as a researcher at the Adirondack Museum and then taught American culture and environmental studies at colleges in Ohio and New York. A Forty-Sixer, he was awarded the Eleanor F. Brown Communication Award by the Adirondack Mountain Club and the Kerr History Prize by the New York Historical Association. He earned a BA in English from Princeton University and a PhD in American civilization from George Washington University. This is his sixth book on Adirondack history and culture.

www.ingramcontent.com/pod-product-compliance
Lightning Source LLC
Chambersburg PA
CBHW030437040825
30551CB00003B/11